MANAGING THE
PROFITABLE
CONSTRUCTION
BUSINESS

MANAGING THE PROFITABLE CONSTRUCTION BUSINESS
The Contractor's Guide to Success and Survival Strategies

Thomas C. Schleifer, Ph.D.
Kenneth T. Sullivan, Ph.D.
John M. Murdough, CPA

Cover Design: C. Wallace
Cover Images: Hardhat © iStockphoto/P_Wei, Tape © iStockphoto/onebluelight,
Calculator © iStockphoto/ LevKing

Published by John Wiley & Sons, Inc., Hoboken, New Jersey
Published simultaneously in Canada

For general information about our other products and services, please contact our Customer Care
Department within the United States at (800) 762-2974, outside the United States at (317) 572-3993
or fax (317) 572-4002.

Wiley publishes in a variety of print and electronic formats and by print-on-demand. Some
material included with standard print versions of this book may not be included in e-books or in
print-on-demand. If this book refers to media such as a CD or DVD that is not included in the
version you purchased, you may download this material at http://booksupport.wiley.com. For more
information about Wiley products, visit www.wiley.com.

ISBN 978-1-118-83694-1 (cloth); ISBN 978-1-118-83704-7 (ebk); ISBN 978-1-118-83713-9 (ebk);
ISBN 978-1-118-85245-3 (ebk)

Printed in the United States of America

SKY10059708_111023

CONTENTS

18   Performance Measurement   237

Appendix: Answer Key for Chapter Review Questions   251

Index   253

FOREWORD

Construction professionals need this book. The construction industry contributes to a significant part of our nation's gross domestic product every year. The industry comes very close to the pure free market system our forefathers envisioned with little barrier to entry but also little barrier to failure. In good times contractors do not have to be the best strategic managers, but this is a cyclical business and in difficult times they have to be strategic, decisive, and unafraid to make tough decisions. Until this text, there has been limited published guidance available on effective strategies and processes on how to prosper in the highly risky construction business.

In this work, Tom Schleifer has given away his "secrets" and has gathered two coauthors who were willing to do the same. I have known Tom Schleifer for many years and over those years I have come to respect his vision of the industry. He has the combination of hands-on experience as a successful contractor and the experience of a long career as a construction consultant, researcher, and educator. In his work as a management consultant to the surety industry, Tom has probably seen and assisted more financially distressed construction businesses than anyone in the industry. This experience gives him a unique perspective on how to manage a construction business successfully. He wrote the seminal book *Construction Contractors' Survival Guide,* which has been acclaimed by thousands of contractors and is a reliable guide to industry credit grantors.

Tom is commonly referred to as a "turn around" expert because of the number of construction companies he has rescued from financial distress. He applied his knowledge and experience as a consultant to the surety industry and later became a sought-after private consultant. Tom is also a sought-after lecturer for industry organizations around the country and in his current role as a research professor at the Del E. Webb School of Construction at Arizona State University, he has been a driving force in developing research and models of the industry that are both revealing and realistic.

Tom is known for telling it like it is. Anyone who wants a crisp clear view of the construction industry and the strategies required for success should definitely read this work.

Terry Lukow
Retired CEO Bond and Financial Products, Travelers

PREFACE

The construction industry continues to become more sophisticated, has fully embraced and implemented computerization, and continues to employ new technologies, particularly in long- and short-range communication equipment. The industry has also moved closer to becoming a commodity, which is the primary reason profit margins are low compared with the historic norms of 10 and 20 years ago. Lower margins increase risk by allowing less room for errors. Profit enhancement over the next decade will result primarily from productivity improvements.

Management decisions alone determine whether an organization will succeed or fail in the construction business. Many construction professionals believe that they lose money or fail because of weather conditions, labor problems, inflation, unexpected rises in interest rates, the high costs of equipment, a tightening or shrinking of the market, or simply bad luck. Actually, none of these are primary reasons for contractor failure. They may contribute to failure once a bad management decision is made, but they are not the basic cause of failure. Failure to obtain a profit or even survive is often attributed to factors or conditions over which management has no control. This is not the case.

Surviving or even thriving in the industry does not imply management can drop their guard. Growth, expansion, and basic ongoing operations usually involve change. And change has risks associated with it, which can make or break an organization.

Unfortunately, past success is not an indicator of future success. When a company expands in size, takes on larger projects, or goes after projects of different kinds or in different territories, it requires management decisions to reduce the risks inherent in such change. An enterprise may be doing fairly well, or even very well; however, the stress to an organization of growth or change can cause a weak component to become a fatal component. Change itself has to be managed to minimize risk.

One problem is that construction professionals don't talk to each other or share information freely enough to create and share a body of knowledge on how to manage a construction business. They meet, converse, socialize, make jokes, and tell stories about each other, friends in the trade, and competitors, but they don't tell each other about how they run their businesses. They don't talk about mistakes that cost profits, postpone deadlines, or worst, jeopardize entire companies. There is an unspoken code among construction professionals, but virtually everyone knows it or learns it soon enough: Learn the essence of the industry, acquire the essential information not taught in classes, and secure knowledge of the business on your own through personal experience. Once gained, that information is yours. You've paid for it. You've earned it. It's your personal property. Let others earn it and learn it the same way. Because of this attitude, the industry reinvents the wheel every day.

There are some very good reasons for the existence of the code, but most are outdated. Construction is a very horizontal industry. If you include those people involved in ancillary jobs or in the manufacturing and transporting of building materials, there are more people involved in producing the built environment in this country than in any other industry. It's a huge industry with over a million individual businesses. The turnover rate as companies go out of business and start-ups replace them is phenomenal. Construction is, undoubtedly, the most highly competitive and high-risk industry in the United States. Consequently, if construction professionals can survive a mistake, correct it, and learn from it, they know something their competitors might not know. This information gives them a competitive edge they are not likely to give away.

To put this in perspective, think of the other industries that compare to construction in terms of their contribution to the gross national product: the automobile, steel, oil, and aircraft industries. All of them have extensive training programs to train their personnel from entry to top-management levels. And these training programs are ongoing and under continual review and revision. These industries have also developed a system of checks and balances on their decisions and strategies; they have boards of directors to ensure accountability and monitor managerial decisions and techniques. In contrast, nine times out of ten the construction professional has learned how to run a construction enterprise by watching, by working various jobs, and by working with someone who has been successful in the past. There is a lot of truth in the old construction story that all you need to start a construction enterprise is a pickup truck, a hard hat, a box of tools, a cast-iron stomach, a forgiving wife, and a bad temper. Many have started with less.

There isn't one way to run a construction enterprise, but there are only a few ways to run one successfully. People in the industry seem to assume that there is no single method to be taught because there are as many ways to run a construction enterprise as there are construction enterprises. The tragedy to the industry is that this is a false assumption. While each business develops its own style, there are only a few ways to run, structure, operate, and manage a construction business successfully and control business risks. Some like to believe that sheer energy, drive, ambition, know-how, and guts will get them through this high-risk industry, and it will for a while. However, there comes a time when that energy and drive have to be organized, given direction, planned, and held to objectives. Without appropriate structure, proper organization, and risk recognition and management, success is elusive.

Put simplistically, a construction enterprise has only three primary functions: getting the work, doing the work, and accounting for the work—marketing, operation, and administration. These three functions are separate and distinct from each other, and to be dealt with effectively they should be analyzed separately. Business and technical activities should be broken down into these functions, and time and energies budgeted to manage the functions appropriately. It is imperative that one person must have direct personal responsibility for each of these three functions. This can mean three different people or two or one. It is not unusual for a small enterprise to have one person handling the getting and accounting for the work while another handles operations. Neither is it unusual for one person to handle all three functions. But as a business grows, the functions remain distinct. The personal responsibility must be clearly recognized. One role may appear more important and significant to the operation of the company. But that is mere appearance and neglecting any of these functions is courting failure. They are each essential to success.

Individuals who accept responsibility for one or more of the three primary functional areas of management described previously are key to the organization whether or not they own a piece of the company. It is critical to success that the individuals believe they are ultimately responsible for the success of their functional areas of the business and accept responsibility personally, not as a functionary or an executive, but as a principal in the organization.

There are reasons for the code, but unfortunately it has some very negative spin-offs. Without cross-fertilization, sharing of essential information, and the collection of a body of knowledge available to the industry as a whole, there is a significant time lag between improvements and modernization within the industry and when that information becomes widely known. Better ways to operate and organize construction enterprises and improve management and production are usually closely guarded secrets for those who discover them. There are methods that can be learned and must be employed to control the risks in this extremely high-risk industry, to allow participants to take informed risks, and to allow managers to learn from the mistakes of others as well as their own. Unfortunately, major business strategy mistakes in the construction industry are often fatal to the enterprise, and until now, no one has collected sufficient data on the subject to provide the hard facts.

This book identifies the common elements of contractor failure and how to avoid them, and presents strategies for construction organizations' business success. This book, a product of the authors' decades of experience in the industry and the collection of a wealth of information on the causes of construction business failure, categorizes the causes and creates a learning tool. Hundreds of actual examples have been annualized to isolate the common causes or elements of failure. One or more of these elements has appeared in every business failure studied. The discovery of what *not* to do leads to researching, developing, and cataloging the elements of successful construction enterprises: what to do.

We begin with organizational principles because many managers believe that quality construction and a few breaks should guarantee business success. They don't. Identifying and understanding the elements of past failures will provide the reader with the means to achieve future success similar to the way generals review past battles and the way medical researchers study illnesses to find preventions and cures. If any good at all can be said to come out of the tremendous number of industry failures over the years, it would be to pass along the lessons learned from their mistakes. This book is dedicated to the fine construction professionals who didn't make it, and the information is offered in the name of those good men and women.

Thomas C. Schleifer, Ph.D.,
Kenneth T. Sullivan, Ph.D.,
John M. Murdough, CPA, MBA
Scottsdale, Arizona
March 2014

ACKNOWLEDGMENTS

The contributions brought to this book by the coauthors, Kenneth T. Sullivan, MBA, Ph.D., and John M. Murdough, CPA, MBA, are new knowledge not understood or recognized until now. Their years of practical experience and hands-on research into how the industry functions will benefit construction professionals and students for years to come. One of the great pleasures of working on this project was numerous stimulating interactions and discussions from which I was able to learn and expand my understanding of many issues impacting the industry.

The tireless work and contribution of Arizona State University Ph.D. candidates Anthony Perrenoud, Brian Lines, and Kristen Hurtado added significantly to the quality and accuracy of this book and cannot be underestimated.

HOW TO USE THIS BOOK

PART 1

After an overview of the industry and a general presentation of the common elements of construction business failure, subsequent chapters address the top five elements. The chapters will define the elements of failure clearly, give real examples, and discuss ways to minimize the risks involved. The recognition of these elements is a tremendous asset. After the elements of failure have been discussed together and separately, succeeding chapters demonstrate how to take these don'ts and use them to develop a positive and competent management attitude and strategy.

PART 2

Chapters 12 to 18 outline proven strategies of success and are an expression of the authors' many years of industry experience and research. Unlike typical management texts this work addresses advanced and proven concepts and methods about how to organize, operate, and succeed in the construction business. It is an executive, CEO, and contractor level how-to manual to understand a complicated and risky industry and be able to navigate and maneuver through it. The authors have developed a profit-centered approach to the high-risk construction business. The second section contains the elements of success: appropriate practices that assist in avoiding and counteracting the elements of failure presented in earlier chapters.

A TEACHING TEXT

Construction schools, colleges and universities can use this book as a teaching text in their construction education programs. To facilitate use as a teaching tool, multiple choice questions are included at the end of each chapter with answers in the back of the book. Also included are critical thinking exercises designed to stimulate group or classroom discussion. There are no answers for these exercises as they are open-ended reasoning tools.

ABOUT THE AUTHORS

Tom Schleifer, Ph.D., joined the construction industry at age 16 and has more than 50 years of contracting and consulting experience. He has Bachelor of Science and Masters of Science degrees in construction management from East Carolina University, and a Ph.D. in construction management from Heriot-Watt University in Edinburgh, Scotland. Dr. Schleifer's experience includes serving as foreman, field superintendent, project manager, and vice president of a construction company that he owned with his brother. From 1976 to 1986 he was the founder and president of the largest international consultancy firm serving the contract surety industry. During this period, he assisted in the resolution or salvage of hundreds of distressed or failed construction firms.

This combination of practical, hands-on experience as a contractor and assisting financially distressed companies has given Dr. Schleifer a unique perspective on the causes of business failure and how to avoid them. Dr. Schleifer, sometimes referred to as a "turn around" expert because of the number of companies that he has rescued from financial distress, advises contractors on organization, structure, and strategic planning while he also writes, lectures, and teaches.

The importance of education in the construction industry is one of Tom Schleifer's favorite themes. He has lectured extensively at universities and professional and trade associations, and authored numerous articles and publications on construction and business management. Dr. Schleifer has been listed in *Who's Who in Finance and Industry*, *Who's Who in America*," and *Who's Who in the World*. He was the 1993 Eminent Scholar of the Del E. Webb School of Construction, Arizona State University.

Publications by Dr. Schleifer include books (*Construction Contractors' Survival Guide*, John Wiley and Sons; *Glossary of Suretyship and Related Terms*, CMA Press), video and audio tapes (*Schleifer's Construction Profit Series*), and a newsletter (*Schleifer's Construction Forecast*).

Kenneth Sullivan, Ph.D., is an associate professor in the School of Sustainable Engineering and the Built Environment at Arizona State University (ASU). Professor Sullivan possesses Bachelor of Science and Master of Science degrees in Civil and Environmental Engineering, an MBA in Real Estate and Urban Economics, and a Ph.D. in Civil and Environmental Engineering, all from the University of Wisconsin—Madison. He grew up in a construction family and has spent his entire life in the industry, with his father working in the transportation infrastructure and heavy civil sectors of the industry, and eventually running his own company.

Professor Sullivan specializes in performance measurement, risk management, best value contracting, organizational transformation, and accountability systems. His research processes have generated over $6 million in research funding and have been implemented in hundreds of real-time projects valued at over $3 billion. In addition to all types of design and construction delivery, he has led project efforts in networking systems, software, campus dining, sports media rights, radio systems, furniture, health insurance, document management, and others.

He has authored over 120 peer reviewed publications and, in addition to teaching several construction management courses at ASU, lectures extensively to industry professionals, universities, and educational seminars across the globe.

John Murdough, CPA, MBA, is a nationally recognized instructor and workshop leader who combines in-depth construction industry experience with a rare gift for teaching. John's course in the business of construction at Arizona State University's Del E. Webb School of Construction and his many seminars are among the most powerful and popular classes available in the construction community. John received the American Subcontractor Association's *Construction Innovator of the Year* award and the *Plus One* award from Arizona State University, recognizing his innovative educational programs.

In John's consulting and accounting practice in Phoenix, Arizona, he is a key element in the success of many contractors of nearly all types and sizes, using his insight and experience in a variety of business and financial areas. He has been involved with the construction industry for more than 35 years, first working for contractors, then serving as a CPA, consultant, teacher, and lecturer for the last 25 years. John founded a construction-specialty accounting and consulting firm in 1989. In addition to financial statement and tax services, John and his associates work with contractors on a wide variety of management and financial issues from start-up through succession planning.

John's additional contributions to the construction industry include serving as an officer and board member for several construction associations and developing and instructing many innovative seminars and courses for contractors. John is one of the developers of the award-winning Leadership Development Forum for Arizona Builders Alliance and is a leading financial instructor for the Sheet Metal and Air Conditioning Contractors' National Association (SMACNA).

John received a Bachelor of Science degree from Pepperdine University and a Master's degree in business administration from Arizona State University. He is a life director at Arizona Builders Alliance.

MANAGING THE PROFITABLE CONSTRUCTION BUSINESS

PART 1

The first section of this book presents an overview of the industry and a general presentation of the common elements of construction business failure. Subsequent chapters address the top five elements in detail with real-life examples including ways to minimize the risks involved. The cautions about what-not-to-do appear first because they can be lethal to even a well-managed company that uses appropriate processes and strategies. The risks must be understood in order to be able to recognize them, avoid them or if necessary maneuver through them. Succeeding chapters demonstrate how to take these don'ts and use them to develop a positive and competent management attitude and strategy.

PART 1

1

MANAGING WITH CONFIDENCE

1.1 LESSONS LEARNED

Successful and failed contractors of similar size and experience work in the same environment. In over 1,000 cases of failed construction companies studied, outside stimuli did not cause failure. Some of the more obvious industry problems like weather, labor problems, inflation, and even a fluctuating marketplace aren't enough to put a construction company out of business. The contractors studied had freedom of choice to manage as they saw fit, and they failed by virtue of making the wrong choices. Most importantly, there are enough cases to see distinct patterns in decisions and other management actions that occurred so regularly they can be identified as common causes of contractor failure. Much can be learned by studying these companies with regard to what not to do, and, consequently, analyzing what should have been done. Understanding, in advance, the risks to success and knowing when they are likely to occur is necessary to avoid or prepare for them.

This study reveals a very disheartening fact: Success in the construction industry, even for very long periods, doesn't guarantee continuing success. In fact, the study indicates clearly that every change in a successful organization, particularly growth, creates a period of risk in spite of all previous successes. Even a carefully thought-out strategic plan cannot eliminate risk, but can only highlight it and alert preparation for it.

Organizations have lifecycles just as living organisms do; they go through the normal struggles and difficulties accompanying each stage of the organization's lifecycle (development) and are faced with the transitional problems of moving to the next phase of development.[1] Management must be skilled at recognizing the change occurring or the symptoms. Organizations learn to deal with these problems by themselves or they develop abnormal "diseases" which stymie growth—problems that usually cannot be resolved without external intervention.[2]

1.2 OBJECTIVES OF THIS BOOK

The good news is that, while the risks are many, they can be dealt with and managed with a proper plan. In Part 1, the primary areas of greatest risk to contractors are identified and analyzed regarding the impact they have and could have on companies. In Part 2, a holistic model is presented that considers accounting and strategic decisions regarding the entire delivery process and business practice of contractors in the pursuit of success.

1.3 MANAGING AREAS OF RISK

Any one of the areas of risk identified, if handled poorly, can create problems for a construction company. If that construction company is already in serious financial difficulty, these areas of risk could lead to putting the company out of business. All changes in a business cost money and perhaps even shake up an organization, but dealing with any one of these areas of risk at a time should not be life threatening to a healthy business even if the change isn't handled well.

Most of the failures observed stemmed from the contractor's attempting to deal with two or more of these risks at one time. Even with good management skills it is extremely difficult to steer a business around several obstacles at once, and each of these areas presents serious obstacles to success. For example, if a contractor, while working on one project, takes a much larger project at a distant site, the degree of difficulty or risk may be easy to see. However, they may be at serious risk if they also have inadequate equipment cost control and poor billing procedures (this compounded risk may not be so evident). Adding more risk areas on top of an already risky decision increases their risk exposure.

Every construction organization will face each of the primary elements of business risk as their business grows and outgrows the organization. This knowledge alone prepares a contractor to face these aspects of growth one at a time, if at all possible. If it is necessary to manage several changes simultaneously, a great deal of authority will need to be delegated, and doing this will test a contractor's managerial maturity.

Consider the following recommendations, based solidly on the experiences of hundreds of construction companies. What these organizations learned can dramatically reduce risk and serve as an example to other construction companies. Following is a summary of the common risks, paired with quick tips to mitigating the effect of these risks. More detail on these risks, case studies, and further analysis are given in the following chapters:

Increase in Project Size (Chapter 3)

    Increase project size gradually.

    Doing a project twice the size of the company's current top size is difficult.

    The risk is proportional to the change.

    Take only one larger project at a time.

Finish the first larger project and evaluate before taking the next one. Don't leapfrog and keep doubling the size of projects.

Unfamiliarity with New Geographic Areas (Chapter 4)

Stake out the best area.

Expand slowly away from it.

Test a new area with a small job.

Head out in only one direction.

Plan carefully for expensive regional offices.

Know when to withdraw.

Moving into New Types of Construction (Chapter 5)

Stick to what the company knows best.

Test a new type of work with a small job.

Experiment with a new type of work close to home.

Complete new work before trying another.

Hire people experienced in the type of work.

Try only one new type of work at a time.

Replacing Key Personnel (Chapter 6)

Decide who is key to the success of the organization.

React quickly to replacement need.

Carefully test replacements.

Train continuously.

Pare down as a safety valve.

Watch for burnout.

Managerial Maturity (Chapter 7)

Management is always tested during growth.

Understand the management needs of a growing company.

Learn to recognize signs of inadequate management.

Lack of managerial maturity can happen to anyone.

Get help in time.

Delegate.

1.4 RECOGNIZING SIGNS OF POTENTIAL TROUBLE

Attending to the following indicators of a possible underlying problem will help keep an organization on track:

Unexplained tight cash flow

Decline in profit margin

Disproportional increase in overhead

Increase in the turnover of personnel

Increase in claims activity

Late accounting information

Changes in accounting information reported

Unexpected borrowing

Increase of internal disputes

Decrease in the quality of work

Too many excuses

Departures in the accounting staff

Inadequate time to do anything well

The presence of one or more of these conditions can have very logical and legitimate explanations and indicate nothing more than a bump in the road. However, they are often signs of more serious organizational problems. These conditions were present in hundreds of companies studied, where management either didn't recognize them or didn't have the management skills to deal with them. Any of these conditions is worth a hard look. Resistance from staff to necessary fact-finding is a sure sign that something is wrong because the problem may be in the department being questioned. When middle or top managers want to keep their areas of responsibility to themselves and resent even a nonjudgmental inquiry, a problem exists. A well-run business fosters an open attitude that encourages everyone in the organization to be well informed and to pull in the same direction. Inquiries by top management should not be viewed with suspicion.

1.5 LAYERS OF MANAGEMENT

Top management of a growing organization can become separated from the mainstream of the organization to such a degree that the indicators of trouble become invisible to them. This can and does happen in organizations of all sizes for various reasons. Sometimes the contractors take so much of the workload upon themselves that there is very little time to manage well and no time to plan, or review progress or performance. Management lacks the time to step back to see how the company is doing, to get the big picture. There is little time to think, and little or no planning takes place because most plans quickly become obsolete. People become high on their own adrenaline and merely react to the rush of activity.[3] It happens with the same frequency in small and large organizations.

Other contractors are unaware of trouble signs because their middle managers don't report them, and top managers of large companies cannot observe all details at all times. Still others are just not observant. They're great at putting construction in place but not skilled at overviewing the business and being alert to subtle changes that are affecting it.

1.6   OWNER VERSUS TOP MANAGEMENT

The fact that in the construction business a company's owner(s) and its top manager(s) are usually one and the same creates a unique management problem. The owner of a construction company is the ultimate party at risk and, as such, should set the goals and objectives of the company. Management from the president on down is responsible for carrying out those goals and to conduct the day-to-day operation of the business in compliance with them. It is the owner's role to see to it that the management accomplishes the goals and objectives of the company.

In most closely held construction companies, no one is playing the important role of the owner. Owner-managers can wear both hats, but not at the same time even though they may believe they are doing just that. Most contractors are entrepreneurs, and entrepreneurs typically experience great difficulty in relinquishing control of their businesses. Some try to change their skills and behavior but many fail. The key to a successful change is for the entrepreneur to recognize that a new stage in the organization's lifecycle has been reached and that the former mode of operation will no longer be effective.[4] All but the very smallest enterprise needs the oversight of its owner even if that is one in the same person as the top manager.

Contractor have a choice: to work alone or get help. If they decide to work alone, they must set aside certain times, three or four times a year, to take off their manager's hat and put on that of the owner. This shouldn't be done during the day-to-day activity or with interruptions to do management work. If the office is selected as the place to do the owner's job, evenings or weekends work best when interruptions are minimized. Even assured of uninterrupted time, it is difficult to force the mind away from the day's pressing management tasks at the office. Partners or minority shareholders should be included to the extent that they understand an owner's role as distinct from their management responsibilities.

1.7   DISCIPLINING PERFORMANCE

The task before the owner is to review corporate goals and to critically scrutinize the company's performance, their own performance, toward accomplishing them. It is checking on self, but it is particularly meaningful if worked at. It is necessary for the owners to forget about how hard they have worked to accomplish their objectives and about everything that has stood in their way, so that they can objectively evaluate their progress. Owners need to ask themselves whether the company is moving in the direction decided upon the last time the owner's hat was worn, and if it isn't, they need to ask for reasons. Everyone needs to be accountable to someone, and if owners choose to be accountable to themselves, they must go through this exercise regularly and make it work. Many fall into the trap of believing that they are doing this every day or on a regular basis. This process must be objective, and that requires discipline and mental distance from daily activities.

A critical performance analysis and the disciplining of the struggle toward goals and objectives are missing ingredients in the formula for success in almost every closely held construction enterprise. This lack of management account-ability is one reason for the high number of business fatalities in the construc-tion industry. It's not only lonely at the top; it doesn't make good business sense to run an entire organization without some independent outside verification of strategies and effectiveness.

1.8   BOARDS OF DIRECTORS

It is very strongly recommended, regardless of size, that contractors have an active board of directors or board of advisors (Figure 1.1). This is not a board in name only, or one that meets for dinner occasionally, nor should it be made up entirely of family members or insiders. A board of directors or advisors rep-resents the owners' or stockholders' interest and is particularly useful where the owners and management are the same people. Just doing what seems to make sense in the day-to-day management of the business may not in fact be good for the company in the long run, and an active board helps keep a company on track.

© iStockphoto.com/ezzystock

Figure 1.1   An Active Board of Directors Helps Keep a Company on Track.

1.9 ACCOUNTABILITY

Board members or advisors from outside the company or industry provide the missing ingredient in the closely held corporation—accountability. They protect the company's assets by asking hard questions. Since they aren't caught up in the day-to-day activity of the business, they are able to see the whole picture.

The formation of a board of directors or advisors to help control risk makes good sense and is not a threat to a contractor's rights as an owner or to their independence. If directors' liability insurance becomes an issue, a board of advisors can be formed which may minimize or eliminate the liability question. Legal advice should be sought in structuring the board to limit exposure to the outside board members. If the idea of more outside members than insiders creates concern, start with a small number, say, five, and have only two outsiders. An odd number of board members is customary to prevent ties, but in actuality there are more discussions than votes in a closely held company, and an even number works as well. What is most important is that the outsiders be truly outsiders, not the corporate accountant, attorney, or banker and not necessarily close friends, unless they are known to be objective enough to participate and speak their minds regardless of the relationship. The only prerequisite is that they are successful business people. They need not be knowledgeable about the industry or the particular business. In fact, it is preferable to have members from an industry other than construction and different from the other members. If outside members are placed on a working board, it will be less effective for family members or inactive board members to continue to sit unless they are qualified business people and can interact as peers.

1.10 SELECTING THE MEMBERS

The right people can be selected from the business community in which the owner lives or works. The size of the construction company should be a guide in selecting candidates with helpful experience. Most business people are complimented by being asked even if they don't have time. Candidates need to know how much time is being asked for and the intention to have a real working board, so they will know their input will be important. If not already known, they need to be interviewed to ensure the compatibility and comfort of both parties.

Professional board members are paid and well worth the expense. For small companies $1,000–$2,000 a meeting may be enough; large companies may spend much more. Some companies pay by the meeting and some by the year plus each meeting. Many companies do not pay additional compensation to inside board members, although some do.

The board should meet as often as the owner feels they need to; however, meeting less than twice a year means the members will lose touch, and meeting more often than once a month means they will become more like management or insiders than advisors. Asking each member to sit on the board for a specific term of two to four years allows the flexibility in membership without the embarrassment of asking someone to leave who is not adding value. Stagger terms so the fewest board members leave in any year. The bylaws of a corporation set out the minimum duties of a board of directors, but members should understand that their main function is to set corporate goals and to evaluate and discipline the achievement of those goals.

An active board of directors or advisors with outside members not only will protect assets and make the business owner's job easier, but also can be a great source of strength during tough times. It's one of few business decisions involving no risk.

1.11 IMPORTANCE OF CREDIT

Construction contractors generally rely on a steady source of credit for their continued success (Figure 1.2). Secured equipment loans and unsecured working capital lines of credit as well as surety credit to secure bid, payment, and per-

Figure 1.2 The Importance of Credit in the Construction Industry Cannot Be Overstated.

formance bonds are the lifeblood of many construction enterprises (particularly those that are growing). Without continued bank and bonding credit, many construction companies would cease to exist very quickly. It is as necessary an ingredient for success as management ability, manpower, and bricks and mortar. Herein lies a major exposure for the contractor in this high-risk business. Continued credit relies on, among other things, favorable performance as demonstrated or delineated in the annual financial statements of the company. Put another way, credit is fairly easier to get when things are going well, but gets more difficult and can even dry up during tough times. One bad year for some contractors may not cause problems in this area, but two or more bad years in a row can restrict or even dry up the credit necessary to continue in business. The risk for the contractor is having internal or external events affect the business such that management cannot reverse losses within one or two fiscal years or before it adversely affects the credit so necessary to the continuance of their business.

> *...credit is easy to get as long as you don't need it...*

In this credit-centered industry, the risk of losing or restricting credit or having it restricted is more critical than the risk of losing money, which of course is usually the cause of losing credit. The idea that credit is easy to get as long as you don't need it is dangerous. A vicious and not well-understood reality is that a loss of credit, not cash, is the real cause of business failure because as long as a company has credit, it can fund its operation. Without it, operations cease.

1.12 VOLUME VERSUS PROFIT

The construction industry is growth oriented and volume driven, as illustrated by the popular belief among contractors that "if you're not growing, you're going backward." Contractors pay far too much attention to their own and their competitors' annual volume and often refer to each other in the context of volume and growth: "There's Tom, he did $5 million last year" or "That's Jim Smith, he went from $50 million last year to $75 million this year." Construction executives should be more concerned with profits; growth is healthy if measured in profit dollars, not simply in volume.

Emphasizing volume and growth may be fine in a good marketplace, but a real problem in a declining one. A contractor's failure to cut overhead and to live with lower volume in a declining marketplace can hurt the business. If a contractor is not organizationally and personally prepared to do less annual volume in a declining marketplace, the organization will need to take greater risks than necessary just to hold volume. They will be trying to get a larger share of a smaller market, and will need to cut margins, go farther for work, seek larger jobs, or do a different type of work. No matter what the plan is to hold volume or grow in a declining market, it dramatically increases business risks in an already high-risk industry.

The alternatives are often not seen because a volume-driven contractor doesn't consider reducing volume or downsizing the business as an option as alternatives to chasing more work. In a declining market, one should expect to get less work, but continue to perform it at a profit. A job may need to be bid at less than normal markups, but bidding at cost or less to hold volume is suicide. It is far less risky to downsize than pursue desperate bidding to maintain size.

A declining marketplace never put a construction company out of business; however, the steps the contractor takes or the business decisions made can change the operation and/or affect profitability to the extent that the business begins to lose money. These actions occur pre-loss or during profitable times. It is often the pending drop in volume that precipitates the decisions. If a contractor fully understands the magnitude of the risk and carefully considers the consequences of failure, cutting volume and overhead is a lot easier to face.

1.13 EMPLOYEE BENEFITS AND COMPENSATION

An area of exposure for the growing contractor is the establishment of unrealistic employee benefits and bonuses during good years. In a business that is inherently subject to peaks and valleys in market conditions and profitable opportunities, a construction enterprise needs a well-thought-out, sensible approach to employee and owner compensation. Some contractors give out unique and expensive benefits, such as country club memberships and expensive cars, usually in good or peak years. Because benefits are easy to give and hard to take back, they need to be realistic and affordable over long periods; otherwise, they backfire in tough times.

Bonuses need to be considered very carefully. Performance bonuses are common in the construction industry and probably overused and abused more than in most industries. A common, costly error is in the awarding of bonuses that are not tied to project profits. I have seen large bonuses given to field people for just bringing a job in on time—a job that ultimately lost money. A more innocuous problem is awarding bonuses for a portion of a job, like certain trade work, if it is properly performed at a profit even when the overall job loses money. From where does the bonus money come? There are far too many jobs in this industry that are salary plus bonus, with the bonus coming every year regardless of the performance of the company.

Bonuses need to be tied to the performance of the entire company. The argument against this is the need for project incentives; however, if some projects generate bonuses—a cash outflow—in a year that the overall company loses money, the loss is increased by the amount of the project bonuses. The argument that the people on those projects did their job doesn't hold water because the people on those projects, like everyone in the organization, are working for continuing employment (among other things). Continuing employment is contingent upon company-wide profitability, which includes all projects—those that did well and those that did not. In addition, the profitable projects may have been less complicated or simply bid correctly, neither of which is owing to anything field management did. Another possibility is that the losing project may have been bid incorrectly or been more complicated, and field management may have actually saved more from reducing loss than the profitable project made.

This is not to say that project incentives should not be used under any circumstances, but that they should be used with discretion and tied, if at all possible or at least partially, to overall company profits. Exclusively using project bonus incentives does not tend to focus all employees on the overall good of the organization and can actually cause unproductive competition for resources that can do more harm than good.

1.14 BORROWING

There is nothing inherently wrong with debt. Many, if not most, construction enterprises borrow for many reasons: working capital, equipment purchases, business acquisition, to leverage equity, to accommodate seasonal or cyclical peaks,[5] to name a few. In fact, some debt in the capital structure of a company is preferred by some accountants to minimize owner risk or improve return on investment.

Uncontrolled or ill-planned borrowing is usually an indication of a problem that increases risks. Uncontrolled borrowing means borrowing without planning or simply borrowing to cover unanticipated operational needs on short notice. When a business needs money that no one expected it to need, management (or the board of directors or advisors) should quickly ask why something is not going according to plan. Well-thought-out business plans anticipate the timing and amount of needed capital, and while plans are not always perfect, unexpected borrowing should always be questioned.

Another kind of uncontrolled borrowing can occur when the accounting department is not supervised by a principal of the company. It is common to have seen contractors delegate the use and negotiation of corporate credit to middle managers, who are not involved with the overall direction and goals of the company. Some of these people can work miracles with banks and can manage to borrow to cover losses for years, preventing the pain of poor performance from being felt in the right places. Borrowing cannot, by itself, put a company out of business; in fact, it can save it for a long time. However, uncontrolled or unplanned borrowing is indicative of a company out of control. When such a company begins to publish poor financial statements, it can quickly lose control of its destiny to its creditors.

1.15 BUSINESS PLANNING

Many in the building process are so involved in the planning of the work that they are distracted from, or lose sight of, the importance of planning for the business entity.

In an informal study of several thousand contractors of all sizes and types throughout the country, less than 40 percent claimed to do any type of formal corporate planning at all, and many did not formalize their planning process with anything in writing. That percentage has dramatically improved in more recent years, but it demonstrates that a lot of companies

> *Long-term business planning is essential to the continuing success of a construction enterprise.*

have not gotten the message: "Long-term business planning is essential to the continuing success of a construction enterprise." Planning is the ultimate risk control (Figure 1.3).

Another informal survey of middle managers generated some disheartening information. One-third of the middle managers questioned indicated that while their company had a formal written business plan, the organization did not follow it closely or the contractor changed direction from the plan without notice. Not following a plan is the same as not having one, although many surveyed said that the annual planning meeting had value even if the plan was not followed.

The most common reason given for not planning is size. Believing either that a smaller business does not benefit from a plan or that planning has little advantage for a mid-size or a larger, well-organized company is a serious business error.

Smaller companies often have limited resources and cannot afford a trial-and-error approach. There are over a million separate contracting companies in the United States, and the largest percentage of them is small businesses. In this tough and competitive industry, many contractors succeed by personally and aggressively driving their business forward. However, the importance of effective and efficient comprehensive corporate planning and its impact on the success or failure rate of construction businesses has been understated within the industry. It has been underestimated by contractors for too long. And while more companies are planning each year, the construction industry cannot count itself among mature, sophisticated industries until that number approaches 100 percent.

© iStockphoto.com/BrianAJackson

Figure 1.3   Business Planning Is Best Conducted away from Or Outside of the Office.

The construction industry is undergoing some dramatic changes, not the least of which is a realization among constructors that the tried-and-true ways of running their contracting businesses will not suffice in this highly competitive and high-risk industry.

There has always been a painful weeding-out process in the construction industry by which companies that don't keep up, fail. Others take their place, and some of them fail. Many assume that this is simply a fact of life in a high-risk, high-stakes business. After many years of working with distressed and failing contracting businesses, we have determined that the causes in most cases were management decisions alone.

The three primary functions of a construction business and how they relate to organizing the business will be described in later chapters in further detail. In brief, they are the following: getting the work, doing the work, and accounting for the work. Once a construction business is broken down into these functions, and time and energies are budgeted to treat them separately and properly, the company is fairly well organized, and able to properly manage the organization (the day-to-day activities of the business).

The next step is to address the responsibility of top management for the long-range goals of the corporation: the well-being and success of the business. Properly managing the important day-to-day marketing, production, and administration areas of the business is not enough to assure success; short-term success in no way implies long- range prosperity.

Without some forecasting and planning, businesses can be driven in the wrong direction. To carry the driving analogy further, contracting businesses have no reverse gear. If someone drives too far in the wrong direction, they cannot simply back up and restart. Once the resources and company is committed in a certain direction, changing that direction drastically can be difficult, expensive, and may be too late.

The process of planning may be defined as deciding in advance what is to be done, when it is to be done, how it is to be done, and who is to do it.[6] Planning is not just an effective defensive tool; however, smaller businesses should embrace corporate planning for that reason alone. In addition to identifying future opportunities and threats to be exploited or avoided, effective planning provides a framework for better decision making throughout the company.

A business plan provides guidance to the managers of a company for making decisions in line with the goals and strategies of the business owners. It helps prevent piecemeal decisions and provides a forum to test the value judgments of decisions makers within the organization. Perhaps the most significant value of the well-organized planning process is the improvement in communications among all levels of management about goals and objectives, strategies to achieve them, and detailed operational plans. Planning is not making future decisions. Planning is concerned with making current decisions in light of their future. It is not what should be done in the future, but rather what should be done now to make desired things happen in the uncertain future. Decisions can only be made in the present. Yet, decisions cannot be made only for the present. Once made, the decisions may have long-term irrevocable consequences.[7]

When properly performed, the planning process creates a communications network within even the largest of companies that gets people excited about what's right for the company and how to achieve it. Planning also addresses an area that is sadly lacking in most small businesses today, that is, the measurement of success. Establishing just a fundamental level of corporate planning in smaller contracting businesses has had profound effects on the outlook, attitude, and performance of employees and business owners alike.

One of the greatest selling points for comprehensive corporate planning is that it allows the contractor to simulate the future on paper. If the simulation doesn't work out, the exercise can be erased and started again. In an exercise, decisions are reversible. Ideas may be tested without committing resources to them or betting the entire company on them. Simulating various business scenarios encourages and permits management to evaluate many alternate courses of action. This could not happen in the real marketplace. Picking the "right" course of action becomes more apparent. There is also the possibility that the larger number of alternatives may produce ideas that would otherwise have been missed. If nothing else, the planning process brings more and better factual information to the table, with which management can make decisions.

The mere ability to experiment with different courses of action without actually committing resources encourages the participants in the process to stretch their creative skills in a safe environment. Models of real-world situations provide an opportunity to test different scenarios and their possible consequences.

While no one can predict the future 100 percent, the probability that certain events will have a predictable cause-and-effect relationship is strong. The more planners know about their business, marketplace, and competition, the greater is the likelihood that they can simulate quite accurately the outcome of certain moves. An individual's comfort and agility in planning increase over time with the experience of applying these planning principles in organizations. Planning can become a formidable tool and a competitive edge.

Corporate planning allows top management to accurately predict new opportunities with greater lead time. With more notice and a predetermined course of action, exploiting new opportunities prior to the competition is much more likely. Another side to this coin, and equally important, is that being better able to look ahead will reveal threats to the business before they arrive unexpectedly.

The days of assuming a business is helpless in the face of market forces are long gone. Contractors in great numbers today are realizing that their businesses need not react only to marketplace-created booms and busts that have plagued the industry for so long. They are embracing an approach that suggests they can determine their future direction with proper, proactive planning. They can be assured that their established objectives are met, or they will at least know the exact reasons why not. It is the major responsibility of the chief executive (contractor) to see that the proper planning system for their company is developed and maintained.[8]

Taken one at a time, the common elements of contractor failures described in the next chapter may seem too subtle to cause a business to go under. But when taken in the context that they were enough to cause hundreds of construction companies to lose money for two or more years, the risks become more apparent.

CHAPTER REVIEW QUESTIONS

1. Which of the following are not guaranteed by success?
   a. Profitability
   b. Continued work
   c. Futures success
   d. The best employees

2. What are common risks for a construction organization?
   a. Increase in project size
   b. Unfamiliarity with geographic area
   c. Moving into new types of work
   d. All of the above

3. The fact that in the construction business a company's owner(s) and its top manager(s) are usually one and the same creates what?
   a. A comfortable situation
   b. A friendly working environment
   c. A unique management problem
   d. None of the above

4. The formation of a board of directors or advisors does what?
   a. Helps control risk
   b. Makes good sense
   c. Is not a threat to contractors' rights as owners or to their independence
   d. All of the above

5. Uncontrolled or ill-planned borrowing is usually an indication of what?
   a. Appropriate management reaction
   b. A problem that increases risks
   c. An immediate need for investment
   d. An error in accounting

CRITICAL THINKING AND DISCUSSION QUESTIONS

1. What are the fundamentals of managing business risk?

2. How can organizations improve their ability to recognize potential trouble?

3. What are the important issues when owner and top management are the same?

4. How would you go about selecting a board of directors or board of advisor members?

5. What are the main issues of volume versus profit?

NOTES

1. Adizes, I., and M. Naiman (1988). *Corporate Lifecycles: How and Why Corporations Grow and Die and What to Do About It* (Vol. 1). Englewood Cliffs, NJ: Prentice Hall.

2. Ibid.

3. Flamholtz, E., and Y. Randle (1990). *Growing Pains: How to Make the Transition from an Entrepreneurship to a Professionally Managed Firm* (2nd ed., p. 11). San Francisco, CA: Jossey-Bass.

4. Ibid. p. 14

5. Hunt, A. (1978). *Corporate Cash Management: Including Electronic Funds Transfer* (p. 51). New York, NY: AMACOM.

6. Steiner, G. A. *Top Management Planning* (Vol. 1, p.7). New York, NY: Macmillan, 1969.

7. Ibid., p. 18.

8. Ibid., p. 95.

2

ELEMENTS OF
CONTRACTOR FAILURE

2.1 CAPITALIZING ON EXPERIENCE

Understanding the reasons why construction businesses lose money is the best
way to prevent unnecessary loss. The investigation and resolution of hundreds of
construction company failures have generated a significant bank of knowledge
on the subject. The events and decisions that precede the failure of a construction business can be categorized and quantified in order to define the most common causes of these failures.

One of the most interesting phenomena
revealed by this study is the fact that the
events and decisions that cause or contribute to a construction business failure take
place during the company's profitable years. To look for the causes within the
difficult years when a company is losing money or breaking even is to study the
result and not the causes. It is easy to be misled in a study of bad years because
losing operations can generate unusual events and decisions even if the contractor is unaware of impending loss.

> *. . . the events and decisions that
> cause . . . business failure take
> place during . . . profitable years.*

The events and decisions that precede a construction company failure take
place during the one to three profitable years prior to the first year of breaking
even or loss. Since many companies struggle through several losing years before
failure, the time frame can be from one to four or more years prior to failure.

A study of the events and decisions that caused hundreds of companies' difficulties identified five recurrent and industry-wide elements of risk to potential
profit or failure. The common elements of business failure are:

1. Increase in project size
2. Unfamiliarity with new geographic areas
3. Moving into new types of construction
4. Changes of key personnel
5. Lack of managerial maturity

This chapter explores each of these briefly, occasionally using very general examples of how these elements affect an organization and its ability to make a profit. In subsequent chapters each is discussed in detail, focusing on identifying and minimizing risks inherent in expanding businesses. All of the decisions concerning these elements are consciously made, and the events are clearly recognizable and usually appear to be routine business occurrences. Many contractors making decisions concerning growth or the necessity to expand into unfamiliar locations or new types of construction do not see them as risky or dangerous; in fact, with proper planning and controls, most of them aren't. There is no suggestion here that a contractor should fear growth or other change. What is expressed is that at least one and usually two or more of these events or decisions preceded the failure of a large number of contractors. There is inherent danger in these elements, and a complete understanding of the risks involved is necessary when encountering them. When two or more of these business changes are undertaken at the same time, they are often lethal.

2.2   INCREASE IN PROJECT SIZE

By far the most common element among contractors who fail is a dramatic increase in the size of projects undertaken. The change to larger projects usually occurs during profitable years; however, problems sometimes develop even before the first of the larger projects is completed. Undertaking larger projects is a natural part of the growth of a construction company; the order of magnitude addressed here, however, is two times or greater than the previous largest project.

The size of a project relative to the size of the company and to the size of its normal or average projects has a definite and direct relationship to profitability. When a construction enterprise is operating at a profit doing a certain average-size project and a certain top-size one, there is absolutely no reason to believe that it will profit if it takes dramatically larger work.

Almost any construction firm can build a project that is two or three times larger than it normally does. If a company can construct $1 million road projects or buildings, it can, in all likelihood, construct $2, $3, or $4 million road projects or buildings. As the size increases, so does the strain on the company's resources and technical abilities; however, within this magnitude it can probably get the job done. But the critical question is this: Can it make a profit?

Making a profit on a job twice the size of a company's previously largest project would be at best unlikely. Making a profit from a job four times greater than the largest ever built would be virtually impossible without both additional resources and a tremendous amount of careful planning—unlikely without outside help. Getting the additional resources required might be possible, but how would contractors with no background on a project of such magnitude determine what resources would be needed? Without previous experience, how could they carefully plan the work? Contractors who normally do top-sized jobs of

$1, $10, or $100 million would be working in an altogether different environment than the one they are equipped for if they took on a $3, $30, or $300 million job.

We'll take a $1 million current top size as an example, but the principles hold true for any project.

Case study: A contractor's largest project is $1 million with two or three major jobs at any given time, say, $600,000, $800,000, and $1 million, and probably a number of smaller jobs in the under-$100,000 range. The company's annual volume is $3 million, and it is generating a comfortable profit margin. When work dries up and backlog approaches zero, the managers go after larger and larger projects. They are able to get a $3 million project, and in their estimation their problems are over for a while.

In fact, their problems may just be beginning. Let's look at the impact on their organization. Previously projects took about a year or less to complete. On the average one of their larger projects started about the time another finished, and a third was at its midpoint. On the project near completion they needed to collect considerable retainage, but the one in the middle stages was generating large monthly payments and the one starting up was about to produce some good cash flow through front loading. By handling jobs in sizes they were accustomed to, which normally were in varying stages, they not only had a reasonable cash flow, but also had the time and resources available to look after all of their small jobs and keep them profitable.

Contrast this with one $3 million job. At first, the front load is terrific, but the retainage mounts fast and within six or eight months will become a higher amount than the company has ever had out on all jobs combined. By the end of the job the amount will be strangling the business, and this project will take longer to finalize than anything ever undertaken. While the project is similar to the work the organization has done, they may be surprised at the level of inspection and supervision they are subjected to by the architect or engineer. Municipal, state, and lender inspections may create more red tape than management is used to or than field staff can effectively handle. Union work rules are usually more strictly adhered to on larger jobs, and security and safety requirements broaden.

The larger project, although similar to other jobs the organization has performed, is not within its experience. The company can get the job done, but making a profit at it is another story. It is similar to a paving contractor who does driveways and parking lots taking on an interstate highway project. Building a parking lot and building a highway are similar but certainly not the same.

2.3   UNFAMILIARITY WITH NEW GEOGRAPHIC AREAS

A change from the geographic area in which a contractor normally works is almost as common an element preceding failure as the change in project size. A contractor's primary area may be one county, half a state, or 3, 5, or 50 states. It is that area in which the organization has normally operated, is comfortable with, and has been profitable (Figure 2.1). While there are many good business

Figure 2.1 A Construction Enterprise Needs to Determine the Geographic Area It Will Work in.

reasons for a company to expand into new geographic areas, such as normal growth, lack of work in primary area, and special opportunities, the risks must be recognized and planned for. Again, the question is not whether the organization can build a similar product in a different location. Rather it is whether a profit can be made at it?

An organization becomes very accustomed to working in an area and can easily assume that their type of work is done the same way everywhere. Yet the differences in customs, methods, procedures, regulations, and labor conditions can be significantly different and expensive if not planned for. Examples include merit shop contractor who bid outside their areas without knowing in advance that the work would have to be performed by a union. In certain areas of the country it is common to construct underground pipe work practically underwater, but in most areas specifications require complete dewatering. In some northeastern states it is almost impossible to keep full crews during the first week of deer hunting season. There are even some areas where local suppliers will give their best prices and service only to local contractors. Regulatory requirements and inspection may differ greatly from an inner city to the suburbs and may be completely reversed when county lines are crossed.

Without going into geological and weather conditions, there are enough potential differences to cause a prudent contractor to want to make certain they know what they are getting into when they take work outside their customary area. Local help, such as a joint-venture partner or new personnel, may be needed to facilitate the project. Compounding the problem, a contractor often takes a distant project that is also much larger than anything they have done in the past because it wouldn't pay to take projects of their normal size so far away.

2.4 MOVING INTO NEW TYPES OF CONSTRUCTION

For a variety of reasons, contractors sometimes change from one type of construction to another or add a new type of work to their existing specialty (Figure 2.2). Companies may change, for example, from highway work to sewage treatment plants; from heavy industrial to tunnel work; from low-rise to high-rise; or from office buildings to hospitals.

The need for research and planning before taking a new type of construction work is well recognized by contractors. What is very often drastically underestimated is the entrance cost—the amount paid for the learning period during which an organization adjusts to performing a new type of construction work. Hiring one senior person who knows the new type of work inside out may not be enough. A company may have to complete one or more losing jobs before it can execute a new type of construction profitably. Unfortunately, some companies do not survive this change.

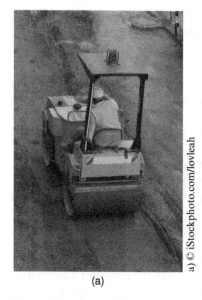

a) © iStockphoto.com/lovleah

b) © iStockphoto.com/Picsfive

(a)   (b)

Figure 2.2 (a) Building a Parking Lot and (b) Building a Highway Are Similar but Not the Same.

Figure 2.3 Most Construction Projects Are Highly Specialized And Many Contractors May Be More Specialized than They Realize.

Most contractors are more specialized than they realize (Figure 2.3). Some construct several types of buildings, for instance, but seem to get mostly one kind. They may call it luck, but it's probably because they are better at bidding and constructing that type of building. Contracting organizations usually start out and remain with types of construction in which they have expertise, and their growth and success are based on the continued perfection of that expertise. Over time they become better able to estimate their kind of work and, therefore, become more competitive at getting it. They also become better at organizing and putting the work in place and become more profitable at it. Being able to plan and execute the construction of a bridge does not mean a person can profitably plan and execute a building.

A more subtle change in type of work is the change from public to private or from private to public sectors; this change, even when the project is a company's normal size and in their own area, has cost numerous firms a great deal of money. Again, it certainly can be done with a healthy respect for the differences and risks involved and some planning. Indeed, many companies do both public and private work and have been doing so profitably for years. There is no suggestion here that it shouldn't be done, but many contractors did not recognize any differences in advance and proceeded to price and produce the work for a loss.

Between public and private work considerable differences exist in:
- Qualifying for selection lists
- The criteria used for selecting the contractor
- The amount of collaboration between contractor, owner, and others
- The quality of work expected and delivered
- The amount of changes assumed to have been allowed for in the bid or to be contracted for separately during construction (change orders)

Qualifying for bid lists works differently in the two sectors. In public work, bidders usually need to prequalify with the public body, the state, or other agencies. However, these lists are open to all contractors, and in most places any contractor can qualify with a little effort. A lot of start-up contractors achieve their growth within the public sector. Their size of project may be restricted at first by bonding requirements, but once they have prequalified, they have a good source of work to bid on because they don't need to "know someone" to bid a public job. This is one of the reasons public jobs usually have many more bidders than private jobs.

Most private sector work, on the other hand, involves selective lists that are more difficult to get on as owners or architects pick the selected contractor, often in a less formal and less transparent manner. Few start-up contractors can find their way onto the better private-sector select lists where the number of proposers is usually fewer than on public projects of similar size. The number of perspective contractors on a project statistically affects the number of projects a company has to go after to get one. This, in turn, impacts the cost of doing business, which affects profit margin.

While some public bodies are required, generally by regulation, to award all work to the low bidder, private-sector selection is usually made with as much concern for quality as for price. The public awarding party has limited control over the bid list or who gets the work. The parties are often strangers, and the award of projects and the administration of them are at arm's length. A public project is usually administered "by the book." The contractor intends to perform according to the specifications and no more. The opposite is true of private work where the awarding party picks the proposers, may or may not open proposals publicly, and often ends up working with a known or at least preselected contractor. The owner, architect, and contractor are much more likely to collaborate on a private project.

Public work is sometimes bid at a lower go-in price than the same private work, and the number of change orders and extras may be greater on public jobs. The reasons are several. The lower price going in on public work allows little leeway to do minor changes at no charge, while on private work, with a team approach, minor changes in the work are often handled informally with no change orders.

On a hard-bid public project, change orders often provide the only profit the job will make. Private work is not priced as tightly because all parties usually understand that a fair markup on the work is expected and numerous nuisance change orders are not. The contractors for private projects need to preserve their relationship with the architect, engineer, and owner for future work. They usually build a reasonable fee and profit into the price, anticipating the necessity for minor changes, and then set about building the project including any incidental changes. A fair profit is earned as a team member.

Very often an organization that does exclusively public work bids too low on a private job. When they begin to go after extras as on public jobs, they run into problems. The architect and owner on private projects, not used to this approach, genuinely think they are being mistreated. The organization has unwittingly created an adversarial situation that too often leads to disputes and claims.

The differences in public and private work are often little known by the players to the extent that constant and avoidable disputes result. The contractor, owner, and architect conduct themselves in what each considers a proper businesslike manner yet the disputes continue. This is probably why private project select lists are here to stay.

2.5 CHANGES IN KEY PERSONNEL

There are three primary functional areas of a construction business, and each must be adequately managed and supervised in a successful contracting enterprise. The primary functional areas are:

- Estimating and sales (getting the work)
- Construction operations (doing the work)
- Administration and accounting (managing the business)

In every successful construction enterprise, a top-level manager is responsible for each of these areas or, in many cases, one person is responsible for all of them or two people share the responsibilities.

If a company is making a profit, it is primarily, if not solely, because of the efforts of these individuals. If one of them leaves, there is by definition no track record of profitability for the new organization. This is a simple reality in business and even more so in the construction business that is so often a closely held small- or medium-sized company.

Some will point to a business with six or eight of the good project managers and say, "That's why this company makes money." But someone may also point to the person who is primarily responsible for construction operations and say that he or she is the reason this company has those six or eight good project managers. The same can be said about two or three key estimators, and some will say the same about the person primarily responsible for getting the work. Successful companies do not relegate responsibility for primary functional areas of their companies to middle management.

The loss of a profit-making top manager puts a construction company at risk; this risk should not be combined with others. The top management team of a construction enterprise is very small compared to other industries because the labor side of the business is field managed like a subcontract, and some contractors even subcontract all fieldwork. The corporate organization is separate and distinct from the field organization. The quality of field management often relies primarily on the quality of the key person or persons responsible for construction operations. If a key person in charge of construction in an organization leaves, the company is permanently changed and at risk until his or her replacement proves that they can do the work for a profit (operations provide the entire cash flow for the company).

On the estimating/sales side of a construction business, one or more key persons will be responsible for the firm's pricing strategy. These managers will usually take a firsthand part in bid preparations and will determine the final price. The takeoff and estimating staff may be a great asset to the company, but the top managers put them together and are ultimately responsible for the success or failure of capturing the work. If one of these people leaves the company, the organization no longer has a proven team that can get the work.

The areas of administration and accounting are much overlooked and underrated by contractors. If there are two top people in the organization who are responsible for the three primary functional areas of the business, one of them will be stuck with the administration and accounting functions; usually these fall to the person responsible for getting the work because sales and estimating are more of an office function than construction operations. In some construction organizations it is difficult to determine who is the top manager in charge of administration and accounting because this function is not often recognized as a primary area important to a company's success. It is often relegated to middle managers even in medium- and large-size companies.

This problem is most acute in growing, medium-sized firms. When the business is small, the contractor runs the entire business, including such details as signing the checks. They are therefore close to the accounting side if only by virtue of paying the bills and having a continuous knowledge of the bank balance. If borrowing is required, they are the ones who explain it to the banker. Administrative needs are few. The small contractor may or may not keep minutes of important meetings, confirm things in writing, or even reply to all of the correspondence received. The small contractor is in continuous communication with the relatively few players on their work in progress, and as a result, the impact on the business of poor paperwork and administration is reduced. As the company grows and the staff increases, administrative and accounting duties are often relegated to middle managers.

If a principal in a construction firm is not responsible for this important primary function, the enterprise is improperly managed. If a dedicated, capable manager who takes personal responsibility for the administrative and accounting functions cannot be identified, the company has a serious problem and is not organized for success. It would be no different from an army marching into battle with no one in charge of its supply line. It's a machine with pieces missing.

Whether this functional area is properly treated by one of the principals or relegated to a manager, it is at risk if the person ultimately responsible for the company's administration and accounting during profitable years is lost to the organization. The accounting staff, under new management, has no track record for monitoring the company's progress with accurate fiscal information.

In summary, one cause of company failure is inadequate replacement of the person or persons responsible for one of the three primary functional areas of the construction enterprise. Typically, the changes in key personnel that contribute to or cause problems take place while the business is profitable.

2.6   LACK OF MANAGERIAL MATURITY IN EXPANDING ORGANIZATIONS

This element of contractor failure is perhaps the most widespread of all. It is very often found in conjunction with one or more of the other elements; in fact, it may be a contributing cause of all the other errors. Many construction organizations are founded by one person. These entrepreneurs who survive the high mortality rate for start-ups usually enter a growth stage. The qualities and abilities required for a contractor to succeed at a small construction business are not necessarily the same as those required for the success of a larger construction business. Confidence and independence, the very traits that cause entrepreneurs to want to be in their own business to begin with, mask for many of them the risks of growth.

Many entrepreneurs assume, "If I succeeded at X volume, I'll do twice or three times as well at two or three times volume X." At some point in the growth of every enterprise, however, the organization must change; it must become more sophisticated. At these junctures, more authority must be delegated, more complex systems and procedures will be required, and more sophisticated people may be needed to handle them. Most entrepreneurs seem to instinctively follow a "command and control" strategy, which was once considered the best approach for a growing business and inherited from the technique developed by Alfred P. Sloan at General Motors during the 1930s. Sloan made the complexity of huge organizations manageable by dividing them into smaller parts, which were in turn subjected to top-down command and control.[1]

Unfortunately, when true delegation is required some command and control must be given up and some founders have great difficulty with that. It would be nice if these changes evolved slowly over the growth period because they would be less drastic and easier for the contractor to digest. But in the real world that isn't the way it works; the contractor can't hire half of a person or put in half of a new system.

Knowing when and how to make organizational changes becomes an aspect of running the business that tests the true entrepreneurial skills of the contractor in a growing firm. The organizational changes necessitated by growth, particularly major reorganizations, need to be made during successful times to assure continued success. The key to success in management is not to eliminate all problems, but to focus on the problems of the present stage of the organization's lifecycle so it can grow and mature to deal with the problems of the next stage.[2]

The contractor who resists change until he has proof of the need for change by having a losing year may have waited too long. Some of the organizational changes required to go from an annual volume of, for example, $1–5 million to $20 million are difficult to recognize and may be difficult for some contractors to accept even if recognized. Delegating responsibility and authority, hiring top managers to supervise long-time associates, friends, or family members, and sharing financial information with more people are a few of the difficult options a growing company may face. In sharp contrast with command and control is today's thinking on "open book" management where each employee in an open book company sees and learns to understand the company's financial information, along with all other numbers that are critical to tracking the business's performance.[3]

The term "managerial maturity" is used here to mean that contractors' managerial abilities must mature as their businesses do. They must change from doing everything themselves to building an organization that can do everything as well, or even better, than they did. Contractors who are unable or unwilling to change their organizations to deal with their growth effectively must either curtail their growth and level off or face the risk of the business outgrowing its own organization. Attempting to do $100 million worth of business with a $20 million organization is suicidal.

CHAPTER REVIEW QUESTIONS

1. A contractor's primary area is what?
   a. That area in which the organization has normally operated
   b. That area they are comfortable with
   c. That area they have been profitable in
   d. All of the above

2. Between public and private work there is what?
   a. Minimal difference
   b. Considerable differences
   c. No difference
   d. Some differences

3. In every successful construction enterprise, a top-level manager is responsible for each of these areas or, in many cases, one person is responsible for all of them or two people share the responsibilities?
   a. Estimating and sales (getting the work)
   b. Construction operations (doing the work)
   c. Administration and accounting (managing the business)
   d. All of the above

4. Typically the changes in key personnel that contribute to or cause problems take place when?
   a. When business is slow
   b. When business is busy
   c. While the business is profitable
   d. At the end of the year

5. When is true delegation required?
   a. When it is easy to accomplish
   b. When some command and control must be given up
   c. When it needs to be resisted
   d. None of the above

CRITICAL THINKING AND DISCUSSION QUESTIONS

1. What are some of the issues when project size is increased?

2. How would you plan to expand into a new geographic area?

3. What are the exposures to an organization that change the type of work they normally do and how can they be overcome?

4. Why is a change in key personnel a serious issue for a small- or mid-size construction enterprise?

5. How would you go about planning for a succession in management of a construction business?

NOTES

1. Dimancescu, D. (1992). *The Seamless Enterprise.* Essex Junction, VT: Omneo (p. 4).

2. Op. Cit., (1988). Adizes *Corporate lifecycles: How and why corporations grow and die and what to do about it* (Vol. 1, p. 32). Englewood Cliffs, NJ: Prentice Hall. (p. 4).

3. Case, J. (1994) *Open-Book Management.* New York, NY: Harper Business (p. 37).

3

INCREASE IN PROJECT SIZE

Chapter 2 briefly describes the common causes of construction business failures. The next several chapters will provide a complete description of the elements of contractor failure, how to avoid them, and actual case studies of contractor failures related to the elements. This chapter concentrates on the most frequent cause of contractor failure, which is undertaking projects that are much larger than a contractor is accustomed to doing. There are times when contractors will feel pressured to take on larger projects, but there are serious reasons that contractor should stick with what has made them successful.

For instance a start-up company will typically grow for a number of years until it establishes a size that the founder(s) are comfortable with or have targeted. In this case, attempting larger and larger projects is required to find and establish a comfortable size. Obviously this is one of the risks related to start-up companies that are breaking into the market. These growing pains are difficult to avoid and are a large part of the reason that start-up construction companies have an extremely high failure rate. Continuing to take on larger projects during a short growth period magnifies the start-up's risks.

A common reason an established firm attempts an enormous project is due to lack of work. A contractor can run out of work for a lot of reasons: declining market, increased competition, higher interest rates, or local construction moratoriums. When backlog is down, larger projects seem a quick way out. If the size jump isn't too great, the risk is probably better than the alternative of completely running out of work. But as the size increases, so do the risks. A project fifty percent larger than any previous one carries much less risk than a project two hundred percent larger.

Failed companies have used all the excuses for taking on the final big project: "A close client wanted us to bid on it"; "The job was right next door to our office"; "We had an in with the owner"; or, the worst one of all, "We adjusted our bid after we were informed of the other bids." Often the reasons for taking on much bigger jobs are opportunistic and the risks are far too great for any of these reasons to make good business sense.

3.1 LIMITS OF GROWTH

One of the main reasons contractors take on jobs larger than previously attempted is because the industry is highly volume driven. Taking on much larger projects to achieve rapid growth or expansion is highly risky. For every business, regardless of the industry, there is a limit to the rate at which it can grow safely. The problem is finding that limit before passing right through it. Organizational resources are stretched to the limit when increasing sales require a seemingly endless increase in people, financing, equipment, operations, and space.[1]

Determining the limits of expansion is not easy because there are few rules or restrictions. In fact, there are some highly respected management specialists who don't believe that there is a limit to expansion. But people should hesitate about extreme growth, because there are enough companies that no longer exist in the industry even though they were household names during their meteoric rise. While critics may point to specific reasons for each of these failures, the fact is that rapid growth itself is dangerous: not always fatal, but always a risk.

There are fundamental financial constraints to healthy and sustainable growth. The management of growth requires careful balancing of sales objectives with the firm's operating efficiency and financial resources. Cases in which companies have overreached themselves at the altar of growth have filled the bankruptcy courts. The trick is to determine what sales growth rate is consistent with the realities of the company and the financial marketplace.[2]

Each entrepreneur has limits to his abilities, available resources, and capital. Each organization is capable of doing just so much. During periods of rapid growth, construction companies change dramatically; they really become new untested organizations with a lot more work to do. The prior organization that was such a success is gone forever. If an organization is to grow, its management must grow. Note that growth does not mean more of the same, growth means to change. The organization's growth must be qualitative, not just quantitative.[3] Qualitative organizational growth takes time and needs to occur prior to sales growth. It usually takes more time than it takes to capture larger projects and, almost universally, construction companies increase management after additional work is on hand, not before. Growth for the sake of growth is risky in any business, but growing in the construction business by taking on projects two to three times larger than anything done before is by far much riskier.[4]

> *Growth for the sake of growth is risky in any business…*

3.2 INCREASED RISKS WITH LARGER PROJECTS

Primarily, the increased risks involved in drastic changes with larger project sizes can be attributed to the lack of experience. An organization with a profitable track record doing projects of a certain size cannot simply assume they can profit from a project of any size (Figure 3.1). Some might disagree and say, "I used to do much smaller jobs than I do now and I'm still making money." This may be true, but how long did the evolution take before becoming profitable? There is a

Figure 3.1   The Size of the Projects a Company Bids Should Be Closely Aligned to Its Profitable Experience with Similar Work. Big Isn't Always Better for a Construction Company.

big difference between having money in the bank and making profit; this will be discussed in a later chapter. A company may be able to double the size of their projects, but that does not mean they should double it again and again. A geometric growth rate might sound impressive, but it typically turns out disastrous. It is difficult for a contractor to know just how large a jump in project size can be attempted at modest risk. However, it is appropriate to understand the nature of the risk and how to evaluate it before staking the business and its future on a giant project (Figure 3.2).

3.3 CASE STUDY

Consider the story of a construction company that averaged annual revenue of sixteen million dollars in commercial and multiresidential work. They had been in business for 15 years and had experienced steady growth from their success. They had been doing two or three major projects a year along with a lot of smaller work in a two-county area of their state. Their average-size project had grown proportionately as the company grew. The largest project to date had been $6.5 million, but they considered anything over $3 million a major project for them. One-third of their volume was small jobs, many under $500,000. Profits were always good, and development in their area was on the upswing.

Figure 3.2   Small Projects, Depending on the Size of the Contractor, May Limit the Risks.

When an out-of-state developer announced plans to put up a luxury condominium in the company's area, the contractor's estimator sent for the plans. When the contractor and estimator saw the size of the project, they almost sent the plans back immediately, but hesitated. The project was enormous for them, they guessed about ten million, but as far as they could tell, they were the only local contractor bidding the job. The other bidders were larger contractors from out of the area or out of state. They felt they had a real competitive advantage and decided to bid the job. They had three major projects underway at the time, but only one was bonded. Since it was almost complete, they were able to get approval from their surety for the bid bond, which was very large for their company.

The design was first class all the way, and as the bid date neared, it became obvious that the project was closer to $12 million, higher than the $10 million initial guessed. They had some difficulty getting their prices together since some of the specialty items were from distant suppliers and sources they had never dealt with before. The size of the electrical and mechanical work on this project precluded the ability of the local subcontractors that they were used to working with. This created some difficulty for the estimator because he was dealing with strangers on some very sophisticated systems and controls. He had to be sure that everything was included but not duplicated. There was a last-minute snag on the bonding, but a hastily arranged meeting overcame that and they submitted their final price at $11.8 million. During this overwhelming period the estimator recalled passing on several small worthwhile projects.

The contractor had taken into consideration the owner's specification requirement of having a full-time project manager and a field superintendent. But, he thought he would not need both of them for three or four months until the project got rolling. However, immediately when the job started, the owner's full-time field representative insisted that both positions be filled from the first day on-site. The company had three key field superintendents running their three major projects, one of which was nearly completed. The contractor put his best superintendent on the job as project manager and another as superintendent. This left one superintendent to run two major projects and finish up the third. The contractor figured he would help out also. When the project manager began to lay out the new job for excavation, he was reminded by the owner's representative that the specifications called for a licensed surveyor to lay out the project. The contractor began to see that they were in an entirely new game.

Once underway, both the contractor and his estimator had to spend what seemed to them an inordinate amount of time at site meetings, updating schedules, and reviewing shop drawings and submittals. The shop drawings were a particular problem because, in the past, the company had never been required to formally sign off on shop drawings. They reasoned, correctly, that if they were going to sign off on all submittal and shop drawings, they had better review them carefully. It took a lot of time, and they eventually hired a draftsman to handle the preliminary review and coordinate the submissions.

The contractors' assumptions that the client would overlook some of the specifications he considered to be excessive also became costly. The contractor was required to hire an on-site project engineer even though he didn't believe he needed one. He also didn't think that he would have to follow the strict emergency and first-aid requirements laid out in the specifications; the owner's representative wasn't as laid back with the requirements.

The payments on the project were very slow, and the contractor became frustrated when he was unable to get through to anyone with authority at the developer's home office. Eventually, after using up all of his line of credit to finance the project and keep the job moving, he threatened to stop the job if he wasn't paid quicker. This quickly got the attention of the developer, and the contractor was invited to the developer's home office. Over an elaborate lunch, a senior member of the firm (who happened to be an attorney) explained that they would do their best to pay him as they paid everyone in their normal course of business. But, the developer made it clear that another threat to stop the job would result in the contractor's termination from the project. They weren't happy with his progress (at this point the project was thirteen days behind schedule). The developer warned him to expedite the work to avoid any disputes over losses the owner might incur if the project was delivered late. The contractor went home and managed to arrange a little larger line of credit and continued to push on.

The contractor's two key people felt they were overworked with the large project, which was outside of their expertise and comfort zone. As the job progressed, the level of activity and number of tradesmen were more than they were used to or were comfortable with. The constant presence of the owner's representative and his staff (which had grown to three people by the height of the project) consumed a lot of the project manager's and superintendent's time, and they became very apprehensive.

The figures on the project started to slip, and the contractor was pushing the project manager pretty hard. There was plenty of other work in the area, and the project manager was genuinely apologetic when he quit. He explained that he simply couldn't handle the pressure. There wasn't a replacement for him in the company and with so much work in the area, little chance to hire from outside. So, the contractor promoted the superintendent to project manager, and a good foreman was made superintendent. The contractor was surprised when the foreman was approved as superintendent by the owner's representative. What the contractor didn't realize was that the owner's representative knew he could get more out of a contractor if the field management was not particularly strong.

By the time the project was two-thirds completed, the contractor knew he had a substantial loss on his hands. He didn't know if his bid was too low because it had not been a public opening (the bid was actually within 1 percent of the second bid). He was not sure if his bank would extend his line of credit further, and his cash flow problems were mounting. These financial problems were compounded because the two major projects underway when he started the big one both had finished poorly. He knew that the poor performance on these projects was the result of taking his best men off the jobs to man the larger project. He had expected his less qualified superintendent to do both projects and close out a third at the same time. The contractor hadn't been able to help the overworked superintendent because he had to spend all his time at meetings and solving problems on the big job.

About this time, the contractor came to another realization; he was in the middle of a construction boom in his area but was running out of work. He was doing fewer small jobs than ever because his estimator hadn't had time to bid them. They'd let the small ones go. Yet these smaller projects had always been profitable and were now sorely missed. The larger job was the real problem. Just after getting the big job a $3 million job was available from one of their best clients. They spent a great amount of time putting a bid together but were shocked when they couldn't get a bond for it and had to pass. In the past this was their averaged size job but now they were only able to get bonding for smaller jobs because they had tied up all of their bond credit on the large project. They needed work badly and were promised additional bonding as soon as their year-end statement was available. Their internal reports revealed exceptionally high receivables from the big job, and the surety was getting nervous.

When the financial statements came out, they weren't good. The disproportional payables dried up most of their cash flow. Tight money had caused the big job to lag. Several subcontractors complained to the owner about nonpayment, and the owner put the bonding company on notice of the payment problem. Finally, once the contractor's credit dried up he couldn't pay his bills and was eventually forced out of business.

3.4 CASE STUDY REVIEW

It cannot be said with certainty that if the contractor had passed up this one big job, he would have been in business today. But given the favorable construction market in his area, he certainly didn't need it. This example is particularly poignant for that reason. The contractor was profitable and positioned in a good marketplace. At the time he bid the larger job he didn't need the work. It was quite natural to look at the circumstances of being the only local bidder and see a competitive advantage. The only problem was the size of the job. The fact that he originally guessed that the job would be about $10 million when it turned out to be $11.8 million indicates that he was outside his realm of experience. His biggest job ever was $3.3 million, and he forgot that when he looked at that one for the first time he guessed it would be $3.8 million. The job was in the contractor's own backyard and involved the type of construction he did best. The only thing out of the ordinary was the size. Being out of his realm of experience he could not foresee the impact on his other work, his cash flow, his bonding capacity, or his profitability. He had no way of knowing that the second bidder was only one percent higher, so that wasn't the problem. He simply did not realize the tremendous risk in taking on a construction project that is substantially larger than anything the organization had done before. There are enough cases similar to this one that signals the devastating risks related to when contractors take on large jobs beyond their experience. The trick is to find the organization's niche in the business; that niche includes the project size they do best.

There is tremendous risk in taking construction projects that are substantially larger than anything an organization has done before.

A construction organization should be cautious of building large organizations with fixed overheads in a fickle marketplace.

3.5 UNDERESTIMATING THE SIZE

A common occurrence in the construction industry is underestimating the size of the job before the job is even estimated. When an organization is bidding work considerably larger than they are used to, there is a tendency to relate the work items to the scale of the work to which the company is accustomed. This is particularly true on work that is not taken off by units such as setting up equipment, cleanup, and so forth. If you are estimating man-hours from your own experience, you may forget or not realize that the equipment on the project is three times as big as normal or that the building, road, or bridge is twice as large as those with which you usually work.

The first precaution in considering much larger projects is to carefully review the bidding process. If the estimator lacks experience with the size of the project, they need to make sure they are not scaling down the project in their minds to conform to their expectations. A utility contractor, whose largest previous job was $2 million, captured a $4.3 million sewage treatment plant renovation project. Six months later, they were awarded a $7.6 million job with the next two bidders at $9.2 and $9.9 million. An extensive study of the entire bid was undertaken by the surety company's consultants with the cooperation of the contractor, and there were no major mistakes or arithmetical errors. Yet when the bid was carefully analyzed and compared to an independent takeoff by a qualified estimator, it was determined, quite to the amazement of the contractor, that almost every line item was low. There wasn't a single sizeable mistake, but literally hundreds of separate items were consistently low. Although it was the same kind of work the contractor always did, no one who prepared the bid, or in the company for that matter, had ever worked on a job this size. They simply scaled the entire job down in their minds to coincide with their experience and expectations. In fact, the scale on the drawings for this large sewage treatment plant was smaller than any scale on which the organization had ever bid. The blueprints, of course, were the usual size since they were photoreduced by the designer, which is not uncommon in this kind of work, and the estimators had properly noted the correct scale in their quantity takeoffs. They simply estimated too low in too many places. The size of the loss put this otherwise successful company out of business. Years of hard work and successful projects were all shot on one job that shouldn't have been attempted in the first place because it was too big for the company. They had been in business for over 40 years.

The second area of concern involves doing the job once you get it. Again, organizations with no experience in larger projects often downscale the job in their minds and believe they can run it with as few people as they typically would their everyday work. They need to determine how many key people this job will tie up, and for how long, and then take a look at what other work they have and how that will be affected. The impact of tying up key people needs to be evaluated as well as the inability to go after additional normal-size work in the marketplace. If there is little work around, of course, that is less of a concern, but someone has to ask the hard questions: Can the key people really do this job and make a profit and do they have any experience with this size work? If additional people have to be hired for the project, the risks grow. The ability and loyalty of new people are untested. If they don't work out and need to be replaced mid-project, another set of problems develops.

A third area of concern is variation in costs that may be unexpected as job size increases significantly. Even unit costs that are normally familiar to a particular contractor may vary considerably in the larger project and it becomes extremely important to use information resources like RSMeans to assist in predicting those costs.

Also remember that the owner's representatives, designers, and inspectors will probably be more used to the size of the project than the company's field management. As seen in the case study presented earlier in this chapter, the project management team got over their head with the technical requirements of the project and quickly became frustrated.

3.6   OWNERS AND RETAINAGE

The risk of undertaking a larger project is increased with an unfamiliar client. It's always important to know something about an owner, but if a company is out of work and desperate, an unknown owner probably won't change the decision to attempt a larger project. It's prudent to find out something about their payment procedures and reputation while determining the impact on the company's cash flow.

Take a realistic, if not pessimistic, look at the length of time the job will take and ensure that you have accurately planned for retainage rates. If there is a reduction in retainage when the job is 50 percent complete, determine whether this is definite or at the owner's option and/or dependent upon owner satisfaction. A retainage reduction cannot be assumed. Remember that holding back retainage is considered prudent by many owners, and if there is any question at all it is much easier not to pay than to pay. Determine the effect if the reduction is not received. If retainage is payable only after final acceptance, take a hard look at how long it will really take to collect even when dealing with a friendly owner. This planning should take place before the bid goes in, not after the project is awarded. Be realistic about how long it will take to collect retainages and the effect on the company's cash flow. Well-managed companies predict how much cash they will need at various times in the future, and then attempt to raise that amount before it is actually required. At the very least they make plans for securing the cash when they need it.[5] Huge outstanding retainages are common among distressed construction enterprises that find themselves at the mercy of their bank lenders or sureties.

3.7   ALLOCATING TIME

Once a larger project is undertaken it is critical to give it the time it deserves as it will represent a significant portion of total volume. However, top management also needs to look carefully at allocating time and not forgetting about the other work on hand. The other projects may be smaller, but they may be and should continue to be the company's bread and butter. An organization seldom wants to lose the profits small jobs generate, and may not be able to afford the losses they could produce from lack of attention.

3.8   ALTERNATIVES TO TAKING ON LARGE PROJECTS

What are the alternatives for a contractor if he is running out of work because of a declining marketplace? The hardest alternative to sell in the construction industry is to do less work. It just isn't in the nature of most contractors to accept such a notion, but it is a very viable alternative. Cutting back overhead and becoming a smaller business to suit a declining marketplace is very realistic. If the entire market in an area is soft, then all of the contractors will be looking for work. Larger contractors who don't usually compete with smaller ones will be going after the smaller work. Bidding within a normal or smaller project size has far less risk than shooting for larger work. The problem, of course, is that an organization's volume will necessarily drop with the smaller projects. It is difficult and unlikely to hold volume with a larger number of smaller jobs because in a soft market there are fewer jobs and more competition. To get these jobs at a profit or at least at the break-even point, an organization can only hope to capture a share of them.

Another alternative would be to expand the company's work area and look for work in their best project size and market niche elsewhere. The risk involved with geographic change is discussed in Chapter 4, but this option can be explored and balanced against the risk of going into a very large project. Unfortunately, most construction businesses aren't very flexible. They aren't set up to expand and

> *A construction organization should have "flexible overhead"...*

compress with the availability of work. A general contractor who was originally a concrete contractor is a good example of being flexible; every time work slowed down, the contractor would take a few concrete subcontracts to keep him busy.

A construction organization should have "flexible overhead" and be cautious of building large organizations with fixed overheads in a fickle marketplace. A portion of overhead needs to be flexible, that is, overhead that is easily removed, like short-term leases on some equipment and some temporary personnel in administrative positions. It is prudent for a construction enterprise to maintain 20–25 percent of their total overhead in a form that it is easily shed. Instead of hiring permanent office staff each time a company expands, it can utilize the services of temporary placement services until the new organization is assured or expands again, maintaining some portion of overhead staff on a nonpermanent basis that is easy to let go if work slows down. The same can be done with vehicles and equipment fleets with a portion on short-term rental rather than purchased, allowing much greater overhead flexibility.

It's the permanent overhead that cannot be easily reduced that requires construction companies to compete for sales in a declining marketplace to the point of desperation. Bidding with little or no profit on the work just to support overhead magnifies risk and in a declining market can be suicide. Some construction executives are convinced overhead can't be let go, particularly people, but also equipment, because it will be needed when the business cycle turns around. It is a compelling argument, but if the company does not survive a down market the argument is defeated.

Business is about recognizing, minimizing, and managing risk. It's not about maintaining size or maintaining potential for a future market. Every construction enterprise must operate in its current market by using past experience to navigate. Large construction projects have substantial risk and are to be avoided if the company lacks the experience.

There seldom will be occasions when, for whatever reason, an organization will determine to undertake a job much larger than anything they have ever done. Hopefully, they will consider all of the alternatives and weigh the risks involved, and if they decide to move forward, develop a course of action and stick to the plan.

> *Business is about recognizing, minimizing, and managing risk.*

3.9 CONCLUSION

We have spent more time in this chapter cautioning against taking on projects substantially larger than the organization has experience with than we have on how to cope with them. This is quite deliberate because the risks are so great that they should be avoided if at all possible. If a company takes all the precautions suggested here, there is still no guarantee they will succeed. Don't believe that a five-story building is like five one-story buildings any more than one $5 million job is like five $1 million jobs. Experience in one project size does not prepare an organization for similar projects twice the size. An organization must learn first how to crawl, then how to walk, then how to run, and finally how to fly. Leave out a step and they may learn how to fall, from a considerable height.

CHAPTER REVIEW QUESTIONS

1. What is considered a big job?
   a. A project two times larger than previously completed
   b. A project five times larger than previously completed
   c. A $50 million project
   d. A $500 million project
2. When is a good time to attempt a project that is double the size of the largest project previously attempted?
   a. Never
   b. During tough times
   c. During prosperous times
   d. During a time in which proportionate resources are available to manage the risk of the project
3. Why did the contractor feel the company would be competitive on the luxury condominium project in the case study?
   a. They were the only local contractor bidding on the project.
   b. They had a great relationship with the developer.
   c. They were willing to do the work to keep the business running.
   d. They were the most well-known contractor available.
4. In the case study, what was the reason that the big project destroyed the contractors company?
   a. The contractor's estimator made a bid that was too low.
   b. The contractor's management was insufficient.

   c. The owner's representative made extreme demands.

   d. The contractor wasn't able to get additional credit to pay bills.

   e. All of the above

5. Flexible overhead would consist of which items(s) (circle all that apply)?

   a. Yearly building lease

   b. Temporary staff

   c. Purchased equipment

   d. Short-term leased equipment

CRITICAL THINKING AND DISCUSSION QUESTIONS

1. Why should a contractor limit the growth of their business?

2. Is it in the nature of contractors to limit growth? Why or why not?

3. What risks are involved with taking on a large project?

4. What does it mean to underestimate the size of the job before the job is even estimated?

5. How can a contractor minimize the risks with taking on a large project?

NOTES

1. Lewis, V. L., and N. Churchill (1983). "The Five Stages of Small Business Growth." Harvard Business Review, 61(3): 30–50.

2. Van Horne, J. C. (1989). *Fundamentals of Financial Management* (7th ed., p. 171). New Jersey: Prentice Hall.

3. Adizes, I., and M. Naiman (1988). *Corporate Lifecycles: How and Why Corporations Grow and Die and What to Do About It* (Vol. 1, p. 32). Englewood Cliffs, NJ: Prentice Hall.

4. Lewis, V. L., and N. Churchill (1983). "The Five Stages of Small Business Growth." Harvard Business Review, 61(3): 30–50.

5. Collier, C. A., and D. A. Halperin (1984). *Construction Funding: Where the Money Comes From* (p.150). New York, NY: John Wiley & Sons.

4

CHANGE IN GEOGRAPHIC LOCATION

The business risk in moving from working in a known geographic area to an unknown one is very real and can be evaluated and measured. A change in geographic area is a significant change in business, far from business as usual. If a contractor has no other alternative, a prudent amount of caution and planning is necessary to minimize the risks involved with the change in location.

4.1 DEFINING "NORMAL AREA"

Business as usual for a construction company is bidding, winning, and completing work in an area that has been profitable. Even though doing work in a normal area will never guarantee a profit on a project, there is greater likelihood that profit can be produced within a familiar area. If the organization has made a profit, perhaps even built a business on doing work in a certain area, then continuing work in the same area involves only modest risk. Contrast this with taking the first job away from the normal area of operations.

A "normal area" doesn't necessarily mean a radius of ten miles or a thousand miles it is better defined by an area of success. For those who do residential remodeling in the suburbs, taking a job five miles across the river in the inner city may be out of their area. Contractors should know what their area is and what it isn't. The distance varies from one company to another, but the premise remains the same: Leaving a typical work area to contract elsewhere means, by definition, working in a place where the organization has no experience contracting for a profit. There is no suggestion here that the job won't get done, but there is a question about profit. Even if a profit is made on an out-of-area job, there are greater risks involved than working at home.

> *Contractors should know what their area is and what it isn't.*

4.2 REASONS FOR CHANGING GEOGRAPHIC AREA

There are many good reasons for expanding a construction business geographically: a desire for growth, the shrinking or drying up of local markets, or

Figure 4.1 Local Conditions Are an Important Element in a Contractor's Costs, And Working in Unfamiliar Areas Can Produce Some Unanticipated Conditions.

the opportunity to follow customers or designers. But, the question of whether to spread out geographically should be considered very carefully and the associated risks should be measured and addressed (Figure 4.1). Numerous contractors jump vast distances, sometimes two or three states, to get work without giving it a second thought. Some even proceeded to expand geographically as if it were the most natural thing in the world. However, many have made this kind of leap with poor, even disastrous results, particularly if they are unprepared organizationally. Typically, companies that are busy developing or expanding their market tend to neglect the development of their operational systems and capabilities. As a firm increases in size, an increasing amount of strain is put on the organization and its systems.[1]

4.3 CASE STUDY: LONG DISTANCE PROJECT

Let's look at a utility contractor who was asked to bid a project 250 miles away from their previous normal area, which included work within a 100-mile radius from their office. The area was remote, and only two other contractors agreed to bid. The contractor knew the design engineers well and was asked by them to participate. The job wasn't needed, and it was at the high end of the size work the contractor did, but the fact that there were only three bidders made it very tempting. Although they had never run a job so far from their shop, supplies, and service lines, they felt they could keep it fairly self-contained and perhaps try some special expediting techniques to shorten the overall time and save on expenses. This would make a neat lump of unexpected profit if they got the job; and if they didn't, fine.

They took the job off in their usual way and, in July, visited the site with one of their project managers. They didn't see anything unusual about the site conditions. In fact, they felt that the remoteness of this farming town would simplify traffic problems and street closings. They asked in the local coffee shop about labor and were told that they could get 30 or 40 people the next day if they wanted them. They could also get quite favorable rates at a local motel if they took several rooms for an extended period. They priced out the job as if it were in their own backyard, adding the equipment transportation costs, mobilization, and all of the costs of keeping their key people on-site. They added a little extra markup for nuisance value and got the job.

The job was to start 14 days after bid opening. This was a little tight on their available equipment, but they could lease a couple of pieces at home to replace what they sent to the job. At bid time, things looked fine. There also appeared to be some available farm equipment in the area that might fill in on grading and seeding, and there seemed to be trucks available locally. There was a snag in the county permits, and one of their people had to travel to the location three times in August to get it straightened out for the friendly design engineers who had invited them to bid. The problem was that the county engineer had no experience with projects this size. The delay meant that the job didn't get started until mid-September. The contractor had figured four months start to finish, but later decided to cut the time in half by working their people and equipment 12 hours a day, 7 days a week, and giving them every other weekend home. The workers would make a lot of overtime, and they would save on subsistence costs by housing their employees for a shorter duration. Since they wanted to do the job in two months, the late start didn't seem like a problem.

Everything started out as planned except a couple of low-boys and tag-along trailers were dearly missed on some existing jobs when equipment needed to be moved. The low-boys and trailers were tied up moving the equipment to the remote site. Then the distance began to take its toll. Everything took a day or two to get out to the site. Downtime for equipment repair parts and needed supplies became a nightmare. There was another big surprise when it became apparent that the crew and equipment were very unwelcome in the town. The project had been planned by a town administration that had been replaced in the last election by a group that ran on a promise to stop the project.

The conservative townspeople had no desire for the sewage collection system with its attendant costs and mess, as they put it. They had used septic systems for years. They didn't want the project. The part-time mayor read the specifications from cover to cover and held the contractor's feet to the fire. Street openings had to be protected with flashing lights and flagmen, even on dirt roads. The mayor further insisted that the lights and flagmen be provided until all paving repair had been completed and accepted. The flagmen were needed four times longer than anticipated to comply with the letter of the contract and to protect streets with a total traffic pattern of 20 vehicles a day.

Many of the details are too painful to elaborate on, but after the design engineers went to bat for the contractors, the town administration threatened to throw the engineers off the project. From this point on there was no salvaging the job. The design engineers also wanted overtime pay for all the inspection time caused by the contractor's schedule. Because of the difficulty of getting supplies, there was a lot of over-ordering. Material quantities ran over by 20 percent. In October, harvest season started and all local labor was lost. Some of it was replaced by borrowing men off existing jobs at home, but there weren't enough to hold the schedule. The specifications said no paving could be done after November 30 and no seeding after October 15. The job was stopped. The contractor left and came back in the spring. Labor costs ran over by almost 200 percent. The total loss on the project exceeded 70 percent of the total contract price. Final payment took two-and-a -half years to collect.

4.4   REVIEW OF THE LONG DISTANCE PROJECT CASE STUDY

If this project had been the only loss during the year, the contractor might have survived the experience. But this remote job impacted the home area work in other unexpected ways. The contractor's plan was to use their best men on the remote job. The project manager was handling three separate contracts close to home when he was sent out of town. Two of them lost money as a result of the changes in field management. Other projects had changes in superintendents in the middle of the work because two good superintendents were transferred to the remote job. And there is no measuring the additional impact of taking the company's four best equipment operators out of the system for three months. Key mechanics, almost all of the service trucks, and even the delivery people were missing at various times during the period. Productivity and morale throughout the organization were worse during this period than anyone in the company could remember. The overall impact resulted in six or seven jobs losing money. Before bidding the remote job, the company was six months into the fiscal year and enjoying their usual profit margin. If the loss from the remote job was excluded from the year-end figures, they were still in a hole. Including the remote job made the year a complete disaster.

There is no implication that a disaster such as this will happen every time a contractor changes their geographic area of operations. But there is risk in moving into unfamiliar territory. Was the farming town story a preventable problem? Should contractors check the political climate everywhere they go? Probably, because when they are in their own backyard, no matter how big that backyard is, they usually know the political climate, and the odds are better that they won't get as many surprises. Distance creates costs. Transportation is expensive, but at least it can be calculated.

Even costs that are normally considered to be predictable for a particular contractor may become unpredictable while working in unfamiliar territory. When bidding or evaluating a project in a different geographic area it is critical that estimators use the best information available to predict costs, and many contractors use resources like RSMeans to provide that information.

The important point is this remote job was handled differently than the contractor's other work. Distant jobs are "different" by definition. If an organization has not worked in an area, they should assume they will encounter different conditions. From the case study the contractor's ability to make profit included moving equipment and workers efficiently from job to job and to have quick maintenance of equipment. But even they did not know how important that was to their success and profitability until the long distance project. The synergy of moving people, equipment, and supplies in their confined normal working area was the backbone of their efficiency and profitability. The distant job, although it was bid like any other, actually wasn't like the previous ones; it was a totally new experience.

There are some things that can be done to minimize and control the risks of doing work outside a normal work area, but all of them put together are not as important as stopping and thinking hard before taking the step. Plenty of contractors have done work out of their area, opened regional offices, and made money. It can be done and has been done. But even the few contractors who succeeded were at enormous risk whether or not they recognized it at the time.

4.5 MANAGING THE RISK WITH LONG DISTANCE PROJECTS

The safest way to expand from a normal work area is to begin at the edges of the organization's existing work area (Figure 4.2). Test the profitability as the first project progresses and find the organization's limits. Perhaps the company can move along an interstate corridor, on one side of a river, or up to a state line before profits begin to drop off. Expanding this way limits the drastic changes that may await a contractor in the totally unfamiliar environment of a more distant location. It is also easier to pull back from the perimeter than from a remote location. Dealing with recalls, guarantees, and maintenance periods required by contract is a lot less expensive at a company's perimeter than at greater distances. Reentry to the area may also be possible after a pullback and some thorough planning. That way, at least the experience is not a total loss.

Hiring locally always makes sense but expect some training and familiarization time if the job is to be performed the same way the company has always managed its work and profited. If a company finds out that they don't do things their way in the new area, then they probably don't belong there.

Relocating a top man for the job from the organization's trusted pool of steady people is probably the safest approach. The organization gets consistency in methods and honesty in reporting. This, of course, will strain the home-territory resources, but a contractor expanding in any manner should realize that this is going to happen. It is one of the risks.

Taking one and only one job at a time in the middle or lower range of the company's normal job size is also a good safety valve, either at the perimeter or at a more distant location. Too many contractors take a large job way from

Figure 4.2 The Safest Way to Expand Geographically Is to Do It Systematically at the Edges of Your Existing Work Area.

home, and before the results of that move are in, they take one or two more. This puts too many resources in an untested environment and raises the stakes on an already high risk. It's just not good business to expand further before the results are in. Knowing a distant job is a greater risk than local work strongly suggests that it be watched closely. Take the pulse of the job regularly. Don't go too far afield if a senior person can't afford to devote the time that the experiment or the job deserves.

4.6 CASE STUDY: REGIONAL OFFICE

Consider a small contractor in business for over twenty years getting into his first sizable commercial job, a $3.5 million strip shopping center in the contractor's hometown. He puts a lot of his time and energy into the bid and then into the job and comes out with a fair profit. During the next two years the same customer awards three more shopping centers, and this contractor gets two of them. His normal work area is half of his state. One of the projects is at the edge of his area and the other is in the next state; each one is over $8 million.

A year later the out-of-state shopping center is completed, and the contractor lands many of the tenant improvements within the center. He decides to lease a small part of the tenant space to open a regional office. He hires a well-recommended local project manager and takes on a few small jobs in the area from the temporary project office. Originally this out-of-state shopping center job was a profit disappointment, but the job had some rock problems early on and weather problems later. But, the contractor thought that the tenant improvements would make up for the loss cost. One of the home-office estimators, who loved to fish in the area, was willing to relocate with his family. After one year the temporary regional office had gotten only $400,000 in work. The company was an unknown commodity in the area, and local competitors took every opportunity to advise their own clients of various rumors they made up about the newcomer. They were given one more year to turn it around. The second year they did $3 million in volume, and everyone felt relieved. Unfortunately, the company was in trouble by the third year and didn't survive the fourth year.

4.7   REVIEW OF THE REGIONAL OFFICE CASE STUDY

The $3.5 million shopping center was a big job for this contractor, and he gave it his personal attention. It was at home, and it made money. The next two were much larger and farther away with one in another state where he had never worked. The only thing the jobs had in common was the developer/owner. The contractor had experienced one good project with this developer; however, the experience is usually good for everyone when a job goes well and makes a profit. At any rate, it was clear from the beginning that these projects weren't going to get as much personal attention as the previous one. Yet no one in the company, not even the contractor, ever correlated his concentrated personal attention with the profitability and success of the first shopping center.

Getting these two big jobs at nearly the same time, however, was cause for celebration, not pessimism within the organization. The decision to hire locally for the out-of-state job was designed to keep expenses down, to deal with the local (new) conditions, and nearly all new subcontractors. The project manager came highly recommended by some good people in the area; however, the company didn't know the project manager who was untested. The rock and weather problems could have happened to anyone. No one even wondered later whether or not the local bidders had anything in their bids for these items.

The new project manager had done some contracting on his own. He said he could get some smaller jobs while they were there for the larger one. No one considered the fact that the field office didn't have any key company or support personnel to do these smaller jobs. Nor did they think it imprudent to allow distractions from a project that was the biggest the company had ever done. The truth of the matter was that the idea of a regional office was on everyone's mind from the moment they got the out-of-state job. Even when it was clear that the job had lost money, they actualized a quick scenario where tenant work would create a breakeven position for the entire adventure. No one questioned that home-office overhead costs associated with the project weren't even charged to it. Not only did they fail to consider the cost of bidding, but the costs of the new communication system, phone bills, payroll service, overnight delivery costs, and so on.

Putting a trusted estimator at the new office was a great idea, and he used the same successful and tested approach he had learned at the home office. But with no market analysis or forecast, the problem of capturing more work was a surprise; the contractor simply planned for the same rate of growth the home office was experiencing with a base of the $8 million from the shopping center project. Consequently, $400,000 in sales the first year didn't even cover regional overhead. The threat of closing the office in an area where the estimator had just settled his family provided sufficient motivation for him to capture more work the second year.

4.8   THE NEED FOR PERSONAL ATTENTION

Similar situations occur so often that it's an embarrassment to the industry. Contractors lose sight of how much effort they put into nurturing and sustaining their companies. They forget how much energy, night work, and worry went into developing a successful operation. And then they put untested people with very little backup in a regional office and expect profitable construction. A regional office is like another construction company. The contractor provides them material resources like money and manpower, and they build. But they can't give them those essential resources: the natural instinct, business sense, drive, and sense of timing that made their company successful. The regional office that succeeds in the construction industry has more than home-office support.[2] It has a leader endowed with most or all of the attributes of the founder.

> *A regional office is like another construction company.*

4.9   OPENING A REGIONAL OFFICE

Opening a regional office can be a good way to get into planned geographic expansion, but it's very risky.[3] It's even riskier when a regional office is opened because of an opportunistic expansion. Examples of opportunistic expansions would be because of one big job or because there is a cyclical opportunity. Opening a regional office because of an opportunistic expansion can be risky simply because it involves less time for planning, testing the market, or testing the decision and it also eliminates the opportunity to evaluate alternative locations.

The absolute key to regional office success is the person assigned to run it. The ideal situation would be for the contractor personally to leave the home office and relocate to the regional office. The home office should be running smoothly and profiting before planning to open a regional office. The talent required at the regional level is the same as that which built the business to begin with. Yet a contractor moving to a regional office is impractical for a lot of reasons, the most important of which is that the "mother lode" requires his or her attention. Basically, setting up a regional office with just a good construction person or administrator to head it up, based on the assumption that it will just be an extension of the home office and it will operate effortlessly, doesn't work. The assumption is false.

A key person in the right location with appropriate home-office support is the basic ingredient for a prudent start-up. And that's exactly what a regional office is, a start-up operation, just like a new contracting entity. As the offshoot of a successful company, it may not need to go through as long a growth and development period, but it should be treated as a start-up operation. If enough effort is donated to it, the new regional office may grow without the missteps and mistakes of the parent office. Remember, however, that regional offices should start small and grow prudently to manage risk. Starting out too big is a problem even with the right ingredients.

In starting regional offices there is the local knowledge factor to contend with even with good research or hiring someone locally to manage the organization. A modest start for at least the first year reduces risk. A company should get some firsthand knowledge of a new area by bidding and doing some work but not so much that the lesson becomes too expensive. It is prudent to get a feel for the area, and allow time to make necessary adjustments.

The best planning efforts for opening a regional office usually underestimate the costs involved, some by a great margin. Invariably the costs will be greater than expected and the income slower to develop than anticipated. Unless an organization has past experiences with opening offices to rely upon, financial contingencies need to be applied liberally in budgeting regional expansion. The real cost and exposure needs to be understood in advance to make certain the company understands the size of the risk involved and can afford the effort. An expanding company might run out of cash not because the situation calls for it, but because management overcommits the organization.[4]

Distance is not as great a factor in locating a regional office as it is with the remote single project because in locating a regional office it is common to follow the marketplace rather than a target of opportunity. "The closer, the better" is a good rule of thumb, but a good market is more important. In fact, a good market is a prudent if not necessary ingredient in a decision to open a regional office. Competition exists in every marketplace so attempting to break into a soft market can be extremely difficult where a strong market may provide room for an expansion effort.

4.10 REGIONAL OFFICE CONTINGENCY PLAN

This brings us to the last important risk control item involved in opening a regional office—a withdrawal plan. Even the most extraordinary turnaround efforts exerted on a regional office can fail. Regional offices have inertia; once set in motion, they continue in motion. Sometimes, even after the best planning in the world, some business decisions simply don't work out; this is why a withdrawal plan is necessary. Withdrawal is a strategy seldom even considered in construction. A withdrawal, or escape, plan starts by determining in advance how long to continue supporting

> *Withdrawal is a strategy seldom even considered in construction.*

a regional office if it is not succeeding. Management must determine, in advance, the accomplishments and milestone dates they will use to measure the success of the regional office. They shouldn't be too optimistic with their expectations.

Financial measurements are the most commonly used when developing measurements of success for regional offices.[5] However, making a profit during the first and even second year may be difficult; in fact, a company may expect to lose money in the first year or two. But the company should determine, in advance, how much of a loss they can afford. A definite amount should be predicted in advance, and if losses exceed the amount, the plan should come under close review. A company may be able to afford the additional losses and continue. If not, they should withdraw immediately. More than two modifications in acceptable losses should signal a complete review of the entry strategy and, possibly, the initiation of the withdrawal plan. If the plan for a regional office doesn't work out, withdrawal before the impact threatens the home office is imperative as a risk control measure.

The company needs to be certain they can afford the cash outflow from the home office that a regional office will require (Figure 4.3). They will also want to look at the impact of a loss on the organization's financial statement and credit. Obviously, cash must be available to meet demands when they arise, and failure

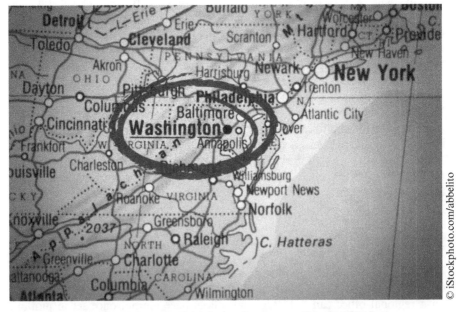

Figure 4.3 A Defined Area in Which the Contractor Works Will Impact Profitability.

to properly estimate the need and provide it will impair a company's credit rating and may even put it out of business.[6] Most companies do not want to capitalize the loss as an investment, so the profit for the entire business will be affected. The bank and bonding company should know in advance of expansion plans, particularly if there is potential or planned losses in the first year or two.

Withdrawal is a strategy seldom even considered in construction industry planning sessions. Yet withdrawal from any strategic business move is a valid, businesslike, and often necessary alternative. The reason it is seldom considered is because so many construction businesses are an extension of the contractor themselves, and there appears to be some stigma or ego impact associated with retreat. There's no room in the business world for this kind of thinking, particularly in a high-risk industry like construction. To control risks, an organization plans and in planning there needs to be alternatives available. Withdrawal is an important alternative; it should be considered as a possibility and included in every strategic decision. Although a contractor might worry about how they will look in the marketplace or to their peers if they withdraw, what's the alternative? Does it make sense to alter a plan to read: "We will stay here at any cost in order to look good to our peers"? Can any business afford that attitude?

Retreat from the market in an orderly, businesslike manner, making sure to tie up the loose ends and leave the area with good public relations in case the company wants to attempt reentry. The truth can be a valuable ally in a withdrawal situation. If the competition was too tough for a regional office to succeed, why not be a class act and compliment peers on the way out. For example, consider a press release like the following: "We find that the fine contractors in this area are well equipped to serve the construction needs of their clients, and our participation in this marketplace, while welcomed, was not needed to an extent that would justify our continued presence." Whatever the reason for withdrawal, it can usually be stated it in a way that neutralizes the rumor mill. At the same time, frankness may gain the organization the respect a smart business decision deserves.

4.11   CONCLUSION

In summary, doing a single project outside a contractor's regular work environment is probably more dangerous than opening a regional office, if only because opening a regional office usually gets a lot more attention. Bidding away from home puts a contractor at a disadvantage to local contractors for many reasons, such as unfamiliar subcontractors and inspectors, labor conditions, geotechnical issues, and so on.

Make sure regional offices are planned carefully not just because there is one enticing opportunity for expansion. Establish milestone indications that, if not met, call for a complete review of the expansion plan. If the expansion proves to generate losses that threaten the home office, implement the withdrawal plan.

CHAPTER REVIEW QUESTIONS

1. What is the "normal area" of a contractor?
   a. Area of success
   b. 100-mile radius
   c. 1,000-mile radius
   d. One or two states away from the headquarters

2. What types of issues are seen when leaving the normal area?
   a. Strain on the organization
   b. Strain on the employees
   c. Unfamiliar conditions
   d. All of the above

3. Which is the best option when bidding on projects out of the area?
   a. Bid low in hopes of establishing relationships in new communities
   b. Bid high to have sufficient contingency in the estimate
   c. Open a regional office to reduce costs of traveling
   d. None of the above

4. What should a contractor do in order to open a regional office?
   a. Understand the real costs
   b. Predetermine performance measurements
   c. Ensure the ability and dedication of the regional office manager
   d. Have a contingency plan
   e. All of the above

5. A withdrawal plan is what?
   a. A no-growth plan
   b. A way to obtain profit outside of your specialty
   c. A mini-business plan that details how the work will be captured and performed and what will be done if the effort is unsuccessful, and that determines in advance how long to continue attempting the new type of work if it is not succeeding
   d. None of the above

CRITICAL THINKING AND DISCUSSION QUESTIONS

1. What is the normal area in your region or in your company?
2. What risks are involved with work outside of your normal area?
3. How could the issues seen in the long distance case study be minimized?
4. How could the issues seen in the regional office case study be minimized?
5. How would you plan and accomplish a retreat from a market you entered and found unprofitable?

NOTES

1. Flamholtz, E., and Y. Randle (1990). *Growing Pains: How to Make the Transition from an Entrepreneurship to a Professionally Managed Firm* (2nd ed., p. 23). San Francisco, CA: Jossey-Bass.

2. Conklin, G. F. (2005). "Risk Management Issues of Branch Engineering Offices." *Journal of Professional Issues in Engineering Education and Practice*, 131(4):284–287.

3. Robinson, Jr., J. H. (1989). "Managing a Branch Office." *Journal of Management in Engineering*, 5(1): 15–22.

4. Adizes, I., and M. Naiman (1988). *Corporate Lifecycles: How and Why Corporations Grow and Die and What to Do About It* (Vol. 1, p. 190). Englewood Cliffs, NJ: Prentice Hall.

5. Tuchman, J. (2006, August 07). "Productivity Benchmarking Effort Produces Results." *Engineering News-Record*, retrieved from http://enr.construction.com/news/finance/archives/060807b.asp

6. Hunt, A. (1978). *Corporate Cash Management: Including Electronic Funds Transfer* (p. 51). New York, NY: AMACOM.

5

CHANGING OR ADDING TO TYPE OF CONSTRUCTION PERFORMED

A contractor seldom changes their business focus entirely from one type of construction work to another. That is, contractors don't switch from building roads and start building only houses. However, expanding into other types of construction work is quite common. It is critical to understand that changing the type of work a construction enterprise performs is an important business event, with serious associated risks.

The reader will notice that this book repeats the following premise: Successful contractors typically become successful by doing a certain type of construction of a certain size, and in a certain area. This can be described as normal work or average projects of a typical type, and success doing one type of construction work is no indication of success in any other type of construction. However, it is not suggested that a contractor should eliminate the idea of expanding into other types of construction. Contractors need to understand that expansion carries certain business risks, great enough to have caused major problems to a large number of successful companies.

5.1 REASONS FOR CHANGES IN TYPE OF WORK

Determining to engage in a different type of work than a firm normally does involves a major strategic decision. Strategic, in this context, refers to the relationship between the company and its environment. Strategic decisions are primarily concerned with factors that are external to the firm, such as the selection of the products-mix that the company will produce. To use an engineering term, the strategic problem is concerned with establishing an "impedance match" between the firm and its environment; in more common terms, it is the problem of deciding what business the firm operates in and what kinds of businesses it will seek to enter in the future.[1]

Comments about diversification made in Chapter 4 warrant repeating here. A number of theories have been developed to explain the reasons for diversification. The reasons for diversification may be linked back to the objectives of the firm, for example, to increase profit, rate of profit, and value of assets, and to decrease turnover. Growth is usually the driving force behind diversification. Growth may be the result of either positive strategic decisions or defensive decisions; however, all diversification moves involve risk and uncertainty.[2]

Sustaining and eventually growing an organization is directly related to the amount of work an organization performs over time. If the reasons for getting into other types of work are uncovered, the most common reason seen is a lack of work in a contractor's marketplace. A close second reason for change in type of work pursued is planned growth, that is, a decision to expand into another type of construction to hasten the growth of the company. Another common reason involves opportunities such as a good client or friend having a job to give out that isn't exactly in the contractor's line of work, but it is perceived as close enough. Whatever the reason for a change in the type of work, the change needs to be approached as an entirely new field of endeavor. The risks of pursuing a new kind of work are the same as those in starting a new business. However, if the risks aren't recognized and addressed, there is an even greater danger of exposing an existing, successful company.

5.2   CHALLENGE: LACK OF EXPERIENCE

The contractor's exposure to risk comes from the same lack of experience discussed in Chapters 3 and 4 dealing with changes in project size and changes in geographic area. The experience problem here may even be greater because a different type of construction may seem like a whole new world for a construction organization (Figure 5.1). What's worse, many contractors don't recognize this and try to perform the new type of work in the same way they do their normal work, with the same equipment and labor force, often without additional training. Competitive success is achieved through people, and the skills of an organization's people are critical. Consequently, one of the most obvious implications of changing the basis of competitive success (by changing type of work) is the growing importance of having a workforce with adequate skills to get the job done correctly and profitability.[3] To get a better feeling for the risks involved, some of the differences in various types of construction are discussed, which illustrate the importance of skills and experience in the particular type of construction being pursued.

A large number of road builders turned to the construction of sewage treatment plants in the past when highway work slowed down or stopped in their areas, some with disastrous results. To make such a switch, a prudent contractor would expect a certain learning curve or wait for their first project to get some experience. The problem is that there is no way to estimate the time or cost required to gain the necessary experience. Circumstances have caused companies to make a move like this without even knowing if they could afford it. What's

Figure 5.1   Some Jobs Are Just More Complex than Others, And Prudent Contractors Stick to What They Know Best.

worse is that there are limited ways to get the experience needed to do the work without first pricing or bidding while having no previous experience in the field.

A common way to overcome the bidding problem is to take on a joint-venture partner. It has been observed that many contractors successfully complete one, two, or several joint-venture projects of a type they were unfamiliar with and at the end still were not able to make a profit when they took on a similar project independently. Thus, being in a joint venture with another company on a construction project doesn't necessarily teach an organization all they need to know to do the new type of work. In fact, if each partner does their share independently, they may learn little or nothing about each other's field.

The road builder looking to build a sewage treatment plan, for example, can likely estimate and do the excavation, site work, and maybe even the concrete work on such a project. Nevertheless, a road building organization has most likely never built round concrete tanks to close tolerances. Furthermore, they would likely not have the necessary experience in the type of pipe work, mechanical systems involved, and the sophisticated control systems required for a sewage treatment plant.

If a company gets over this lack of experience at the bid stage, they are still left with challenges they may face during the construction stage. Simply coordinating the shop drawings and managing the interfaces and physical space problems for all of the systems in a treatment plant creates nightmares for experienced people.

There are numerous cases wherein building contractors and road contractors were not able to finish new types of jobs they were inexperienced in without help, and were unsuccessful in securing a profit. A contractor's success in pursuing work that they are inexperienced with requires a large amount of preparation, planning, and, at times, luck.

> *A contractor's success in pursuing work that they are inexperienced with requires a large amount of preparation, planning, and, at times, luck.*

5.3 CHALLENGE: DIFFERENCES THAT APPEAR SUBTLE

There are differences in types of construction that are even more subtle and more difficult to recognize without having performed the work before. Many building contractors look at all building projects as pretty much the same, until they take their first hospital job. The lessons learned on the first project of a new type are usually severe. They find out that these projects refuse to get done on time, are impossible to schedule, and require them to put three feet of mechanical systems in a two-foot space above the ceiling. The difference between a hospital and a residential high-rise is similar to the difference between an icebox and a refrigerator. While the icebox and refrigerator seem similar on the surface (both keep food cool), a refrigerator has more complexity (i.e., refrigeration systems, mechanical design, etc.) than an icebox (requires simply ice). However, there is not a hierarchy of skills, from what may appear to be a less complicated type of construction to a more complicated type. Each type of construction is different and experience with one type of construction does not imply ability in any other, even seemingly similar, types of construction. Different contractors specialize in each type, with some organizations being experienced in multiple types of construction.

> *Each type of construction is different and experience with one type of construction does not imply ability in any other, even seemingly similar, types of construction.*

5.4 RESOLUTION: KNOW YOUR SPECIALTY

There is no attempt here to suggest that one type of construction is more difficult than another or to suggest that a contractor needs to be smarter or more skilled to do one than the other. The reality is, simply, that every successful contractor is a specialist. By definition, the type of work they have been successful at is their specialty, and these specialties are far narrower than expected. In some cases, an organization may even have more than one specialty and should recognize the distinction.

A construction organization grows and becomes successful by perfecting the skills and abilities needed for their specialty. The organization learns the nuances and subtleties to the point that they can do the work better and bid it better than their competition. Construction professionals often fail to realize how specialized their field is. Contractors may understand the entire industry and how it works and fits together, but when they contract to put work in place in return for a fair profit, they are safe only within their range of experience.

It is imperative that construction organizations determine and understand their particular specialty. Too many organizations, especially those growing rapidly, don't know where their real expertise lies, where their experience is strong or marginal, or the type of work that contributes the most to their success (Figure 5.3). Almost imperceptible changes in the type of work undertaken can make a big difference in the potential for a contractor's profit and success.

> *A construction organization grows and becomes successful by perfecting the skills and abilities needed for their specialty.*

5.5 BACKGROUND TO CASE STUDIES

In this chapter, specific examples of drastic changes in type of work are not used because there are far too many combinations. Instead, the examples used are more subtle changes in the type of work. However, it has been observed that plumbing contractors take HVAC (heating, ventilating, and air-conditioning) contracts, electrical contractors bid and get building contracts, and road builders take dredging contracts. There is a long list of contractors who took work completely out of their field and lost money. There is an even longer list of contractors who took work in their own general field, but outside of their field of expertise or specialty, and lost money. Some went back to their own type of work, while others didn't survive the experience.

5.6 CASE STUDY 1

Consider a sewage contractor in business for ten years and doing about $15 million a year profitably when they had their first losing year. They were losing money on their two largest jobs, which were both within their normal, top-size range and within their usual work area. The jobs appeared to be straightforward underground sewage collection projects, and the contractor pointed to a string of jobs they said were similar in which they had made money. A closer review of their previous projects, however, revealed something that surprised the contractor. Just about all of their successful projects in the past had been precast concrete pipe gravity sewage jobs, including all of their large projects. Several previous projects had ductile iron pipe force mains in them, but these jobs had not profited nearly as well as the exclusively gravity sewer projects. There had also been two small exclusively force-main projects some years back, one breaking even and the other losing money.

The work this ten-year-old company originally bid in its first six years was only gravity sewage work and in recent years was gravity system, force-main, or a combination of the two. The contractor had started and was busy building a successful construction business and hadn't taken note of the fact that they weren't very successful at doing force-main work. They had not been successful at bidding it either. When their financial difficulties came to light, the two biggest projects that were losing money were both force-main jobs and several other losing projects were combination force-main and gravity systems. The estimator reported that because they had missed several force-main bids, they had lowered their force-main unit prices to get the present work.

The problems continued, and the contractor did not survive the numerous losing projects even though all of the exclusively gravity sewerage projects were profitable. This contractor and their people called the company a "sewer contracting firm," not even a utility contracting firm. It may not be a common or even acceptable name, but by definition this firm was a "gravity sewage contractor," in that for their ten-year history they had never made a profit on force-main work or any other type of construction work other than gravity sewage projects. They never recognized their specialty.

5.7   CASE STUDY 2

Not knowing exactly what their real success was based on also caused serious problems for an old-line company. A very largecale, 50-year-old mechanical contracting firm was operating with three divisions just before the company got into difficulties. They had an HVAC division with their own sheet metal shop, a mechanical division with their own pipe shop, and a service division that also did small contract jobs and provided contract maintenance services. The service division had grown significantly over the years, but because the other two divisions had the greater volumes, service was given little of top management's attention. The service division also added efficiency and profit to the other two divisions because it did all of the start-ups, punch-list work, and callbacks for the other divisions. This made sense because the service division had dozens of fully equipped service trucks manned by highly skilled mechanics.

The HVAC and mechanical divisions made more money than the service business, although not as a percentage of sales, and were taking larger and larger contracts. Several years after a succession of top management, the new managers decided that since the HVAC and mechanical divisions represented 80 percent of the company's volume, the service business would be sold. They determined that the sheet metal and pipe shops were turning out products that could be bought at the same cost on the open market, and because the sheet metal shop and pipe shop were located with the service division on expensive real estate, they, too, were sold.

The new streamlined organization now had its estimators and contract managers located in an office park with no shop overhead. They easily replaced the volume represented by the service division, and shops sold and increased HVAC and mechanical division sales 20 percent the following year. Unfortunately, the company never had another profitable year in its remaining business life.

The current management had not been around when the company was growing. They understood marketing very well and had inherited good pricing methods. What they didn't realize was that the service company absorbed most of the start-up, punch-list, and callback costs for the other divisions, allowing the HVAC and mechanical divisions to appear more profitable. The service division did this work for the other divisions at its actual cost, which was much lower than the other division's costs because the service division used lower cost, nonunion labor.

Top management also did not understand that while the sheet metal and pipe shops produced work for the HVAC and mechanical divisions at the same cost the company might pay on the open market, they produced it on schedule. More importantly, changes and modifications were handled quickly and problems solved overnight. The company historically took a large share of work in a very price competitive market and made a profit because it controlled its work by being its own supplier.

With all the problem solvers from the service division gone, the company could no longer produce the work for its former competitive prices. The real problem was that top management did not understand the true expertise of the organization or the basis of its past success. The company got into trouble doing the same type of work they had been doing for many years, only in a different way (Figure 5.2). In fact, they had no idea what their success or specialty was based on (Figure 5.3).

Figure 5.2 The Risk in Change in Size And Type of Project are Compounded When a Contractor's Largest Job Ever Is with a New Type of Owner.

Figure 5.3 Moving from Paving for Private Development to a Public Highway Project Can Be Challenging And Risky.

5.8 EXAMPLE: UNION VERSUS OPEN SHOP

There are a number of other things that can make what appears to be the same type of work actually quite different. These other factors are different enough to create problems and losses if the contractor lacks the experience to deal with them. Take, for example, the differences between operating with a union or open-shop labor force, which can create problems and cost money when attempting to switch from one to the other.

Changing from union to open-shop work or vice versa is a culture shock of its own. If problems are anticipated, most contractors can survive the temporary or permanent switch, but it can still cost a great deal of money. Open-shop contractors taking on their first union job usually choke on the restrictive work rules. This is particularly true on smaller jobs and is not recommended at any price for minor projects. Union contractors doing their first open-shop job may have difficulty finding the skilled labor they had grown accustomed to with trade unions. The reasons for switching can be varied and sometimes personal; however, the business risk should be carefully measured before attempting to switch. Perhaps the greatest risk is the potential distraction from other work that the organization may suffer if it is tied up learning a new way of doing business.

5.9 KNOW THE RISKS

Some may think that the risks involved in making changes in the type of construction are perfectly obvious. Yet, good contractors make such changes regularly and often with disastrous results. A contractor wouldn't deliberately risk their entire business on one project, but the extent and seriousness of the risks involved are not always obvious. If the risks from drastic changes are often overlooked, it is understandable that the risks involved in more subtle changes are seldom even considered. Subtle changes within the type of construction are nearly as dangerous as drastic changes, but are often undertaken with such confidence that construction enterprises are surprised, even shocked, when the roof falls in.

In the construction business a lot of money is turned over, and only a very small portion of it stays with the company, so it is imperative that an organization understands exactly where their expertise lies and how it operates to generate profits. Managers need to know what type of construction the organization does best, and even which subcategories of work they do better, before they can move forward with any work with confidence and reduced risk. Once management has done this, they can decide for themselves how much additional risk they want to take on as they expand and grow the company. If there isn't enough of the company's type of work available to satisfy the organization's appetite, management needs to be very careful about going after a type of work they have no direct experience in and embracing risks they may not be able to measure or even recognize.

5.10 VOLUME VERSUS PROFIT ALTERNATIVE

As previously mentioned, the primary reason given for undertaking a different type of project than an organization is experienced with is due to a lack of the company's primary type of work. However, when volume levels off or even drops slightly because the market slows down and an organization continues to perform only the type of work they are best at, profits often grow without any increase in volume or risk. The more a company produces of their specialty, the better they get at it. Real profit growth, as opposed to volume growth, comes when an organization levels out or slows down slightly and everyone has enough time to devote to the work on hand, including bidding and administration (Figure 5.4). When managers are not rushed or overworked by growth, they have time to make improvements in the work they perform. When people have time to fine-tune their skills in the organization's specialty, there is tremendous profit potential available.

Most will agree that making more money on less volume with little risk is real growth. Delaying a decision to attempt a different type of work because the market slows down is a valid alternative. If a company, as forced by the market, operates at no growth or slightly reduced volume for a while before expanding

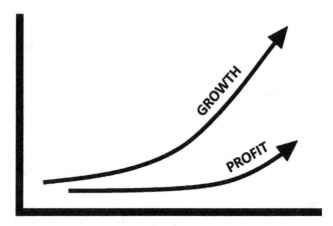

Figure 5.4   Real Profit Growth Should Be Considered before Volume Growth.

into other types of work or bigger jobs, they will at least have greater cash reserves and more experience when, or if, they decide to take on the greater risks of expanding into a type of work the organization has little or no experience in.

5.11 WITHDRAWAL PLAN

As discussed in Chapter 4, a primary element of risk control is planning and having a withdrawal plan. Before attempting any type of work the organization has no experience in, a mini-business plan should be developed detailing how the work will be captured and performed and what will be done if the effort is unsuccessful. A withdrawal or escape plan starts by determining in advance how long to continue attempting the new type of work if it is not succeeding.

Management must determine, in advance, the accomplishments and milestones they will use to measure the success of the effort. There is no need to be too optimistic. Making a profit during the first or second attempt may not be possible and, as stated about regional office planning in Chapter 4, profit isn't the only measure of success. The company may be trying to penetrate a market or be in on the ground floor of a developing or expanding type of work. It is reasonable to expect to lose money on the first job or two, but management should determine, in advance, how much of a loss the company can afford. They will also want to look at the impact of a loss on the organization's financial statement and credit. Obviously, cash must be available to meet demands when they arise and failure to properly estimate the need and provide it will impair a company's credit rating and may even put it out of business.[4] The potential and recommendations for withdrawal should be included in every strategic decision. Withdrawal from performing a type of work can usually be accomplished with few people, if anyone, outside the company taking notice.

5.12   CONCLUSION

If management is convinced they must move into different types of work to maintain volume or support growth appetites, they should evaluate all of the risks first and proceed with caution. It is advised to test the water with a smaller job to minimize exposure. If that doesn't work, the organization can withdraw gracefully (in accordance with their withdrawal plan) and try something else. In any case it is never prudent to bet the whole company by attempting a larger, first-time type of work.

CHAPTER REVIEW QUESTIONS

1. Some of the reasons why contractors choose to change or add to the type of work they perform are what?
   a. Diversification
   b. Growth
   c. Profit
   d. All of the above

2. What are potential ways of getting around a company's lack of experience?
   a. Joint venture
   b. Bid on a project without experience
   c. Obtain a workforce with the required skills
   d. All of the above

3. How can company know their specialty?
   a. Identify where their experience is strong or marginal, or the type of work that contributes the most to their success and allows them to do the work better than any other, at a profit.
   b. Look at their core values.
   c. If they are growing, they are specialized in that field.
   d. A company can develop their specialty overnight.

4. Different types of construction are what?
   a. They are pretty much all the same when it comes to building a project.
   b. They can be performed by almost anyone with basic skills.
   c. They don't require a large amount of sophisticated management.
   d. They are different, and experience with one type of construction does not imply ability in any other, even seemingly similar, types of construction.

5. A withdrawal plan is what?
   a. A no-growth plan
   b. A way to obtain profit outside of your specialty
   c. A mini-business plan that details how the work will be captured and performed and what will be done if the effort is unsuccessful, and that determines in advance how long to continue attempting the new type of work if it is not succeeding
   d. None of the above

CRITICAL THINKING AND DISCUSSION QUESTIONS

1. Regarding the risk of changing or adding to the type of work performed, why is this risk so prevalent among contractors?

2. What could have been done differently in the case studies to produce a more favorable outcome?

3. What were some of the proposed strategies to avoid the risk of changing or adding to the type of work performed?

4. What are some other potential mitigation strategies that could be used to avoid or mitigate this risk?

5. What could potentially be some of the signals that changing or adding to the type of work performed may not be favorable to the organization?

NOTES

1. Ansoff, H. I. (1965). *Corporate Strategy: Business Policy for Growth and Expansion* (p. 5). New York, NY: McGraw-Hill.

2. Hillebrandt, P. M., and J. Cannon (1989). *The Management of Construction Firms: Aspects of Theory* (p.31). Portland, OR: International Specialized Book Services Inc.

3. Pfeffer, J. (1995). *Competitive Advantage through People: Unleashing the Power of the Work Force* (p.16). Boston, MA: Harvard Business Press.

4. Hunt, A. (1978). *Corporate Cash Management: Including Electronic Funds Transfer* (p. 11). New York, NY: AMACOM.

6

REPLACE KEY PERSONNEL

6.1 IDENTIFYING KEY PEOPLE

Most of the built environment in the United States is put in place by small- and mid-size construction enterprises, and the majority of all construction companies are closely held businesses managed by some or all of the owners. The term "key people" can be described as those critical to the success of the enterprise, and as such, are not easily replaced (Figure 6.1). The key people in the majority of smaller construction companies are therefore usually easy to find—they are the owners. In mid-size and larger companies the key people are the owners and closely associated top managers that may or may not be minority stockholders. For many construction enterprises there may be as few as three key people, and for smaller-sized businesses there may only be one or two.

> *The term "key people" can be described as those critical to the success of the enterprise, and as such, are not easily replaced.*

This chapter may not be very popular with middle managers (who are a very important part of any organization), as it concentrates on the one, two, or several people without whom a company cannot function. The definition of these key people are included in the definition of a contractor used here, as anyone who is personally responsible for one of the three primary functional areas of a construction organization (described next) whether or not they own a piece of the business.

The three primary functional areas of a construction enterprise were introduced in Chapter 2 as estimating and sales (getting the work), construction operations (doing the work), and administration and accounting (managing the business). In each successful construction organization, someone has direct personal responsibility for each of these functions. Invariably, most organizational problems in a construction organization that can be encountered can usually be traced to a person in charge of one of these functional areas. A weakness in any of these areas can single-handedly cause a construction business to fail. The loss of the key person who is personally responsible for just one of these areas can and has caused many companies to flounder.

Figure 6.1   Each Construction Company Has Key People Who Are the Primary Reason for the Company's Success, And the Training of Future Leadership Is Critical to the Continued Success of the Organization.

6.2   PARTNERS

Many partners who founded and ran a construction company successfully found that when they split up, they were not nearly as successful as they were together. Very often, it takes more than one person at the top for a construction business to flourish because many people are only good at their own specialty. If one partner is excellent at construction operations, they may fail without their former partner available to get the work and take care of the office. The opposite is equally true. A person who can get the work may need a partner to do the work. They can, of course, hire someone to do any of these jobs, but typical entrepreneurs tend to be doers rather than managers and most haven't had the formal management training to manage staff.[1] It's difficult, if not impossible, to find someone to do a task as diligently and conscientiously as a partner would do the task. Simply sharing ownership with an employee doesn't necessarily transform an employee into a contractor. In fact, most employees who have what it takes to be a contractor already have their own plans for achieving that position.

To develop this point further, in a two-partner breakup, the ideal approach would be to fall back to a business no greater than half the size of what the partners did together as a simple matter of risk control. Neither individual is likely to have experience managing the original-size business alone. Thus, to attempt to operate on the original scale is to assume the missing partner did nothing, or whatever they did, the remaining partner can do in the spare time. The 50 percent solution may be impractical as few contractors would voluntarily accept a 50 percent reduction in volume under any circumstances, and cutting a business back by half may actually cause more problems than it cures. However, limiting growth for a couple of years would definitely be prudent to determine if success continues in the absence of one of the key players.

6.3 FOUNDERS AND SUCCESSION

The loss of a business founder is always a difficult blow to an organization. Few construction companies are patterned after any particular succession model. The organizations are fashioned and shaped by the contractor as they progress and grow. Since many sizeable construction companies were started by relatively young people, it's not unusual to find 30- and 40-year-old companies still being managed by the founder. These contractors often struggled early on to find what it was they did best; then over time, they developed a way of doing it that worked for them.

As a company grows, it often changes how it operates or at least appears to, but the founder's values and approach are usually interwoven into the fiber of the organization. The values and business methodology of a founder are the real reasons successful companies overcome the low survival rate for construction start-ups and are able to grow and prosper. After many years and the addition of numerous managers, these values are often all but hidden. However, they can usually be found embedded in the operating strategy of the business, thanks to the continued participation of the founder. Unfortunately, these values are often seen by the people who will assume control in the future as roadblocks to the company's further growth or expansion and are often described as antiquated or old-fashioned. When the successors take over, the values change too much and lose the formula for the company's success.

This does not imply that companies can't succeed and prosper following the loss or succession of a founder. However, it is an unfortunate reality that many construction enterprises have not survived succession, demonstrated by the very high succession failure rate in the industry. Most of the failed businesses could have survived if they had developed a proactive, appropriate plan for the transition including a healthy, objective understanding of what made the company a success. Many construction companies owe their prosperity to the strengths of one or two key people at the top and do not survive their departure because of a misconception of what made the company successful.

> *. . . organizations are fashioned and shaped by the contractor as they progress and grow.*

6.4   INACTIVE FOUNDERS

A less obvious variation on the preceding scenario occurs when a founder becomes inactive in an organization, and everything continues smoothly for years until the founder dies. A founder can influence a company long after they become inactive. Often, the loyalty of long-time employees prevents them from leaving the new management as long as the founder is alive. Their influence, therefore, helps keep alive old values in spite of new management. In other cases, the new management doesn't have the strength to change things too drastically while the founder remains in the background. One of the most unfortunate scenarios seen is when the retired founder accomplishes a successful management transition, only to be destroyed as the result of negative influence by his estate after his death.

Obviously, when the contractor or active partner is no longer around, a company is more susceptible to changes. It has been observed that large, established businesses approach such a transition with no more planning than a retirement dinner. Some top managers in construction companies and even stockholders believe that a successful enterprise can continue to be successful by inertia. Actually, it has been observed that, as few as one to three key people can drive a construction company. If one of them is gone, the company is in jeopardy until a replacement is found and tested. Even a replacement with equal skills and talents will lack the missing key person's contacts and connections as well as their loyalties. These intangibles are lost to the company and are part of the reason that key people are so hard to replace. It will take a year or more to determine whether or not a replacement will work. The exposure and risk to the business of these transitions cannot be underestimated.

6.5   SUCCESSION CASE STUDY

Keep the three functional areas in mind as the following case is described, their importance to a construction organization, and the fact that a change in the person primarily responsible for any of these areas puts the company at risk and should be treated with great caution. Consider a successful, second-generation, heavy industrial construction company doing about $200 million a year that found itself in serious difficulty after the loss of its chief financial officer (CFO). After a successful transition following the loss of the founder (which included some substantial changes in ownership), the company was financially solid and working in a growth market. The successor's chief executive officer (CEO) was a good leader and was picked by the founder for his conservative approach to the business. He was also chosen to offset a partner who was a wizard in the field, but lacked good business sense.

The two partners had built the business together from nothing and had decided not long after they started to bring in a third partner to run the accounting department. While the new partner wasn't given equal ownership right away, they wanted someone with an ownership interest in this critical area because they had some close calls in the past with bad figures during a tough growth period.

Even before the transition, one of the partners had been pushing for the company to buy a firm that did much of the design engineering on their jobs. The design firm was not very profitable. Consequently, the partner thought they could buy it at a reasonable price and increase their competitive edge by expediting the design of their own work. The other partner liked the idea, but a stumbling block was that their only option was a cash deal. Since their debts were extensive, the CFO (original accounting partner) was violently against the deal. He said that using all available cash and borrowing power, even in a good market, would leave them at the mercy of their creditors for three or four years because the debt service would consume 50 percent of their current annual profits.

The CEO, who by now was a full partner, refused to take the risk, but the pressure to purchase continued, and the issue finally had to be settled by the board of directors. The board decided against the purchase, and the decision really turned on the CFO's financial presentation. The CFO was well respected by all concerned and had an excellent understanding of the business, particularly when it came to cash flow. He could remember both the good and the bad years and understood that with a cyclical, high-risk business like construction, a modest cash reserve and some unused credit were not a luxury, but a necessity. He kept constant watch over every financial aspect of their business and reported to the board that, while they were financially sound and in a growth market, increased competition was eroding their profit margins.

The company's primary market was construction and renovation of industrial facilities for the auto and steel industries. While there was plenty of work around, road building and commercial construction had slowed in the area, causing an increase in competition from other contractors and a decrease in profit margin.

Increasing competition was not the primary reason the board found the purchase too risky. What tipped the balance were graphs showing the impact on the company's cash flow eight years prior when their markets declined, following a dip in the national economy and a lengthy steel strike. The company had remained marginally profitable during the period in question, but their cash flow had turned critically negative from suspended jobs, where retainage could not be collected, and from slow-paying clients. In addition, numerous canceled projects added to their profit problems, but what no one realized was that the chief accountant had saved the day by insisting that overhead be cut immediately. His charts showed that without their reserves of cash and credit at the time, they could have gone under.

Two years after the board rejected the purchase of the design firm, the CFO had a serious illness and was unable to work again. The company faced another serious succession exposure that ended when they hired an accountant from their auditing firm who was very familiar with their finances. Everyone was happy with the selection, and since the marketplace was still strong, the future looked excellent. Six months later, the partner in charge of construction again proposed the purchase of the design firm. The firm was still available, and the partner felt the company needed to make the purchase now more than ever to improve their profit margins in the field.

The other partner was skeptical and again it went to the board of directors for a decision. The partner in charge of construction presented a good argument for the purchase, claiming a minimum 2 percent increase in field productivity based on better design service and faster deliveries of design modifications. The new CFO reported that he had studied the figures given to him by the field people as well as the purchase price information from the design firm. He thought they could definitely afford the purchase. When asked if the purchase would tie up all their cash and borrowing power, he reported that it would be a simple matter to increase their line of credit. The purchase was approved on the condition that a certain size increase in the line of credit would be negotiated before the purchase. No one asked how the new accountant had made his calculations.

As it turns out, the new accountant had taken the average profits for the company over the previous two years and calculated an increase on the basis of 2 percent across the board. This increase was reported to the board as being more than enough justification to purchase the design company. Consequently, the purchase was approved. No one questioned the fact that the design firm was involved in only one-third of the work that the company performed. The board didn't know that the calculations presented were based on an increase in profits for the entire company, and not just an increase in efficiency on the work associated with the design firm.

After the purchase, it was discovered that the design firm was even less profitable than originally thought and some additional cash had to be put into it. However, the increase in their efficiency did increase profit margins on the work involved even greater than originally projected, so the purchase was considered a success. Two years later, one of the company's major steel industry clients became insolvent. This loss in work cut the design firm's volume almost in half. At the same time, the auto industry was slowing down because of import pressures and reduced car sales. In less than six months, the construction company saw the cancellation of three major projects they were to start and had five capital improvement projects that were in construction stopped without notice. The company's volume was cut by a third, and they were suddenly losing money daily. In spite of valiant attempts to cut overhead and find work in other areas, they could not reverse the trend. Their next financial statement was a disaster, and their cash flow made it impossible to pay their debt service, let alone reduce principal on the outstanding loan.

The new CFO's attempts to renegotiate and restructure their loans failed when the bank wanted a written commitment from the construction company's surety carrier that all necessary bonding would continue to be available to the company. The bonding company said that each project would be looked at independently. Filing for reorganization was being considered, when a buyer for the company was found. The value of the partners' stock was practically worthless, and the engineering company was given back to the original owners at no cost. A change in one key person in this company had cost the partners their business because the key person was the only one who really understood the long-range measure of risk. Until the original CFO left the company because of serious illness, he prevented the partners from making the ultimate business mistake—taking on any single risk that had the potential to jeopardize the entire business.

6.6 NEW MANAGEMENT TEAM

For all intents and purposes, changes in the key personnel of a construction enterprise create an entirely new company. The reality is that a formally successful construction company with a new management team has yet to complete its first year of profitable operation. The new management team may very well be successful, and there is no implied assumption that they won't do well. However, no assumption can be inferred that it will be successful. The best that can be said of the new organization is that it is untested. Too often it is assumed that each person in an organization works independently, wherein each person is expected to carry their own weight. However, the reality is that in a relatively small organization, such as a closely held construction enterprise, the functional areas of responsibility are interdependent as are the people responsible for them, and a weakness in one area affects all of them.

> *. . . changes in the key personnel of a construction enterprise create an entirely new company.*

The definition of team is "any group organized to work together."[2] A new person added to an existing management team drastically and dramatically alters the team regardless of the size of the team. It may be for better, worse, or neutral, but the entire team is changed. A more specific definition of "team" from the book, *Wisdom of Teams,* sheds more light on the impact of changing a member of the team: "A team is a small number of people with complementary skills who are committed to a common purpose, performance goals and approach for which they hold themselves accountable."[3] The key phrases are "small number," "complementary skills," "common purpose," "goals," and "approach." The likelihood of all of these matching when there remain few members from the original team is slight. If you substitute an ingredient in a cake mix you still get a cake, but not the one defined in original the recipe. Most people can think of their own sports team analogy. It is undeniable that one person changes a team's potential.

To reduce the exposure of an untested organization a reduction in volume, at least not growing for a year or more, allows time for the new management team to prove they can work successfully together. It is prudent to at least restrain any growth until the new organization completes at least one year of profitable construction operations together.

6.7 ADDING KEY PERSONNEL

The loss of a key person is not the only exposure in replacing key personnel. As a company grows, it must continually add to its management staff and, in rapid or sustained growth, new people are regularly placed in key roles. The company now has a newer, larger, untested organization, and they are performing a greater amount of work. The original team of key individuals may have proven itself, but the new larger group has not yet operated as a team and proven profitable. Few growing companies take the time to analyze their situation after adding key personnel because it is difficult to isolate specific variables when the entire company will likely be occupied with the new growth in various facets. If problems develop some years down the road, there will likely be new people or circumstances on which to blame.

6.8   MANAGEMENT "DILUTION"

A potentially more serious problem a growing construction enterprise will face, particularly during periods of rapid growth, can be referred to as a change in key personnel by "dilution." As a construction company expands, it usually takes on larger and more sophisticated projects, which often requires additional managers and some with more or different experience than existing team members. Hiring new people into an organization and making them managers over long-term employees very often creates problems and jealousies. Groups are important at the psychological level because individuals' actions, thoughts, and emotions can't be understood without taking into consideration the groups they belong to and the groups that surround them (Figure 6.2).[4]

Contractors find themselves determining to grow and expand their businesses because of, and with, the loyal team that worked so hard to generate the current success. Contractors then find that the company outgrows the original team's capabilities and new managers with more or different experience are required to maintain the growth. Existing managers are often offended or feel their years of loyalty are being repaid by bringing in someone above them, and worse, at a higher salary. There are various drivers for choosing to work for a certain sized company such as atmosphere and regular access to the contractor (boss). There is a certain amount of what may loosely be referred to as "hero worship" in the construction industry, whereby charismatic entrepreneurs are able to generate tremendous loyalty and attract people that will work extreme hours and put in staggering effort just to work with the contractor.

As the company grows, contact with the contractor diminishes and new managers are placed between the contractor and loyal hardworking employees, creating discontent, and many leave. Experience shows that a company that grows at a rate that doubles the organization's volume in as little as three years will lose at least half of its core group of key employees as a result of the growth. The negative impact felt by the contractor who does not expect or understand it may be stronger than in a non-construction company. From the standpoint of an entrepreneurial organization, it is clear that something inevitably will be lost as the organization makes the transition to professional management. However, something will also be gained. Just as a plant that has outgrown its container must be transplanted if it is to continue to grow and develop properly,

© iStockphoto.com/gemenacom

Figure 6.2 Efficiency Is Impacted When Too Many New People Are Brought into an Organization Rapidly.

an organization that has outgrown its infrastructure and style of management (and people) must also make a transformation. Failure to do so will lead to a variety of other problems.[5] The problem of dilution of key personnel during rapid growth probably cannot be avoided.

6.9 SUMMARY

The loss or addition of key people creates, by definition, a new and untested management team and puts the company immediately at risk. The risk is often unavoidable, but, when recognized, can be controlled by avoiding the assumption that all is well that looks well. Construction is a difficult business at best and requires a unique person or group of people to construct for a profit, and any change in that group is like starting over. The matching, or team, theory exists here also. People who have succeeded separately or in different groups cannot automatically be assimilated successfully into a new group. Each new grouping has its own set of challenges in matching skills and, therefore, needs to prove its ability to work together and make a profit before doubling and redoubling the organization's bet.

CHAPTER REVIEW QUESTIONS

1. Who are the key people in a construction organization?
   a. The entire organization
   b. The project managers
   c. Only the founder
   d. The one to three people personally responsible for one of the three primary functional areas of a construction organization: estimating and sales (getting the work), construction operations (doing the work), and administration and accounting (managing the business)

2. When they separate, partners often find what?
   a. They can do the same amount of work and can easily hire someone to do the other partner's work.
   b. Replacing a partner is difficult and exposes the organization to risk until it is determined if the new partner has the best interests of the organization in mind.
   c. There is no longer need for a partnership; one can do the work of the two.
   d. They can hire an external management company to look out for their best interests.

3. Creating a proactive, appropriate succession plan for the transition from one key person to another involves what?
   a. Obtaining a healthy, objective understanding of what made the company a success
   b. Understanding the risks involved in operations
   c. Analyzing the financial position of the organization
   d. All of the above

4. What are some of the main reasons for replacing key personnel?
   a. Partners move on or leave
   b. Founders pass away
   c. Growth of the organization
   d. All of the above

5. How can "management dilution" best be described?
   a. Contact with the contractor diminishing and new managers being placed between the contractor and loyal, hardworking employees as a company grows
   b. Adding more businesses and key functions to the company
   c. Losing the original core values of the company
   d. Management taking a less active role in caring about their employees

CRITICAL THINKING AND DISCUSSION QUESTIONS

1. Regarding the replacing key personnel, why is this risk so prevalent among contractors?

2. What could have been done differently in the case studies to produce a more favorable outcome?

3. What were some of the proposed strategies to avoid the risk of replacing key personnel?

4. What are some other potential mitigation strategies that could be used to avoid or mitigate this risk?

5. What could potentially be some of the signals that replacing of key personnel is being performed and may not be favorable to the organization?

NOTES

1. Flamholtz, E., and Y. Randle (1990). *Growing Pains: How to Make the Transition from an Entrepreneurship to a Professionally Managed Firm* (2nd ed., p. 13). San Francisco, CA: Jossey-Bass.

2. Davies, P. (Ed.) (1976). *The American Heritage Dictionary of the English Language.* New York, NY: Dell Publishing.

3. Katzenbach, J. R., and D. K. Smith (1992). *The Wisdom of Teams: Creating the High-Performance Organization.* (p. 45). Boston, MA: Harvard Business Press.

4. Forsyth, D. R. (1990). *Group Dynamics.* (2nd ed., p. 6). Belmont, CA: Brooks/Cole Publishing Company

5. Flamholtz, E., Y. Randle (1990). *Growing Pains: How to Make the Transition from an Entrepreneurship to a Professionally Managed Firm.* (2nd ed., p. 47). San Francisco, CA: Jossey-Bass.

7

MANAGERIAL MATURITY

7.1 START-UP CONSTRUCTION COMPANIES

The exciting business of construction attracts many new start-up companies each year and has been doing so for a long time. Even the largest construction companies in the country today were at one time start-ups and were oftentimes started by a single person.

Attesting to this is the fact that most construction companies carry the name or initials of the founder. Most founders probably never expected or envisioned that their initial efforts would develop into the big and successful enterprises that exist today. In fact, some of today's biggest firms were started by men or women who couldn't find gainful employment and by individuals who simply wanted to be their own bosses. Very few of these contractors started out with clearly defined long-range plans to become nationwide or multinational construction companies in a predetermined number of years. Interviews with numerous founders of large U.S. construction companies offer an insight into their modest expectations at the time of company start-up and their genuine surprise at the size their firms ultimately grew to. The modest initial expectations can present managerial challenges as the company grows in size. Different managerial skills are needed to maintain success as a construction company grows, and contractors must be cautious that their growth does not overreach their managerial capabilities.

This leads to the curious question: "If large construction enterprises were for the most part unplanned, where do big construction companies come from?" The answer in most cases is evolution—they evolve from smaller organizations rather than being carefully planned. To be sure, most large- and medium-size construction companies today have sophisticated planning strategies with elaborate systems and are managed by highly competent businesspeople. But that's today. Almost all of these successful enterprises can point to a time in their history when their longest-range plan was how to make payroll the following week, or the most sophisticated system they used was the back of an envelope. What's more, there are hundreds of failures for each of these start-up successes. Despite its significance in the economy as the second largest industry in the United States, the construction industry suffers from the second highest mortality rate (the food service industry is first) with the average construction company lasting only five years.[1]

7.2   IMPORTANCE OF MANAGEMENT SKILLS

So why do some start-ups succeed and grow into large, long-lasting companies while so many more fail? And why does this pattern continue today? No doubt luck has some effect, but so many failed contractors were competent constructors who worked in industry conditions identical to their more successful competitors, so we can eliminate luck as a primary cause. And since the best businesspeople and entrepreneurs seem to create their own luck, we need to look further for the cause of success.

The only discriminating variable existing between the successful construction businesses and the early and midterm failures is management skill. Management skills are not just more sophisticated leadership skills, which most entrepreneurs have in abundance. Different analytical skills and thought processes are involved as well as the patience to sacrifice shortcuts for long-term optimum results. Management skills include a certain amount of vision, so that planning for and even dreaming of the future can take place. Because start-up contractors may not be trained in management science or inherently gifted managers, they need to develop the mature managerial skills to enable their companies to grow beyond the start-up stage (Figure 7.1).

Start-up contractors find out very quickly that there is a lot more to running their business than putting the work in place in the field. They need to attract and capture new work, account for the money, administer all the details

Figure 7.1   The Managers of a Growing Construction Organization Need to Continue Their Professional Development to Meet Future Challenges.

associated with subcontractors and ordering materials, meet the payroll, acquire business insurance and bonding capacity, and so on. Some start-ups do not continue because the principal never anticipated all of this responsibility, or they simply do not want to do it. Some principals try to remain in the business but ignore the details. Others address the details but not well. This last group makes up the huge number of start-ups that don't go beyond the first five years in the contracting business. The single most important reason these companies do not succeed long-term is a lack of management skills.

7.3   COMPANY GROWTH PHASES

Construction contractors that succeed beyond the early mortality years usually go into a growth phase that can last for the lifetime of the business. Few reach the six- or eight-year point and level off. Most construction businesses attempt to grow quite rapidly during the early stage of their development, and management skills are perhaps the most important ingredient in a contractor's formula for success. If a business remains stable in size, it can be operated in the particularly proven and successful style of the founder. But when a company grows, its management approach must also mature. If a business is expanding even at a modest rate of 10 or 15 percent a year, it will periodically grow out of its own systems and procedures, which in turn dictates that its management needs will change.

> *... when a company grows, its management approach must also mature.*

The senior management of rapidly growing entrepreneurial companies must simultaneously cope with endless day-to-day problems and keep an eye on the future strategic direction. Furthermore, the managers of most such companies are going through the process of building a company for the first time. This is about as easy as navigating uncharted waters in a leaky rowboat with an inexperienced crew while surrounded by sharks. The sea is unfamiliar, the boat is clumsy, the skills needed are not readily apparent or not fully developed, and there is a constant reminder of the high cost of error in judgment.[2]

If a construction company's growth rate is above the 15 percent mark, the need for management development is compounded. When growth is this rapidly paced, management must not only manage the company's operations but also the growth process itself. Planning for and handling growth is an important factor in an expanding business, and dealing with changing plans becomes a process that must be managed. In spite of this reality, many contractors double and redouble their growth year after year with little attention to organizational needs. Because of the stress placed on a poorly managed organization, getting more work than planned for can be as much of a problem for a contractor as not getting enough work.

Management ability is critical not only to the survival and continuance of a start-up enterprise; it is crucial during any growth period in a construction company's lifespan. Many struggling businesses claim good or bad luck affects their success, but it has never been proven to be a make-it or break-it factor. Other factors that are commonly blamed for failure include working in a good or bad marketplace as well as having a good or bad labor pool. Yet these factors do not correlate to construction company failure because every other construction organization operates under the same conditions in a given area.

7.4 LIMIT OF MANAGERIAL EFFECTIVENESS

Company growth presents a wide range of challenges for construction contractors. Construction contractors running fast-growing companies often find it difficult to know when they have reached the limit of their managerial effectiveness (Figure 7.2). Medium-size companies often lack someone in the management team who is vocal, sophisticated, or objective enough to warn the management team of a problem. For other contractors, the biggest obstacle is an inability to delegate authority within their management teams. Strong

© iStockphoto.com/pzAxe

Figure 7.2 There Can Be Too Much to Juggle When Managerial Maturity Is Exceeded.

entrepreneurial contractors in large companies often surround themselves with weak top managers because they won't give anyone real authority within the organization; yet many of these same contractors then wonder why they are working so hard.

The number of contractors who have enjoyed many years of success only to fail because they let their companies get larger than they could effectively manage is staggering. This happens when contractors' businesses outstrip their managerial ability, or when the contractor is not able or wise enough to bring in the new management talent that the organization needs. A high level of managerial and personal maturity is required to know and admit when one has reached the limit of one's own effectiveness. If a company cannot adapt to changing circumstances, it will either fold or drop back to a marginally surviving company.[3]

7.5 COMPANY GROWTH AND MANAGEMENT THRESHOLDS

The issues of resource allocation, personnel, planning, and systems gradually increase in importance as a company progresses from start-up through slow initial growth to more rapid growth. These resources must be acquired somewhat in advance of the growth stage so that they are in place when needed.[4] Yet the point at which growth will outstrip management ability is not easily predicted because the threshold is different for each contractor and for each management team. There are no specific limits for an operation of a given size or type. The contractors must therefore be able to identify when company growth is outpacing their managerial maturity. The best judges of this are the contractors and their management teams, who are usually too busy to look for signs of growing pains. Unfortunately, the easiest way for a company to identify if they have outstripped their management ability is by retrospective analysis; that is, only after they have already approached or exceeded their real limits.

> *... contractors must therefore be able to identify when company growth is outpacing their managerial maturity.*

7.6 TELLTALE SIGNS OF INSUFFICIENT MANAGERIAL MATURITY

It is critically important that contractors are aware of the risk of overextending themselves beyond their managerial threshold and to continually be on the lookout for telltale signs that change is needed. An organization that is managed by more than one person has a slight edge in that several observers provide a better system of checks and balances. Managers are able to observe and critique one another's methods and results, which tends to bring management shortcomings to the surface faster.

There are several telltale signs that a construction company is trying to manage more than its leadership team is actually capable of. For example, a series of internal gripes or small problems, most often concerning personnel issues, signal possible trouble. Especially disturbing are complaints from long-time trusted employees, an increased turnover among middle managers and field forces, an increase in owner and designer complaints, a drop-off in the performance or response of trusted subcontractors, an increase in job site accidents, and increased absenteeism. All of these factors tend to grow proportionately with the growth of the business. Yet when several of these challenges occur at once, and for no apparent reason, it is usually a sign that something is wrong. When management reaches a point of reacting to business pressures and concerns as opposed to acting on new issues and planning, continued growth is a mistake. Without management improvement, expansion could break the organization.

7.7   THE CHALLENGE OF MANAGEMENT CHANGES

When a construction organization doubles or triples in size, it is no longer the same company it was before growth, and it usually will not be successful if it maintains pregrowth management methods. Yet changing existing management practices is extremely difficult for some contractors and for good reason. One of the more difficult changes is to evolve from the close-knit, team-oriented camaraderie of a small company into to a mid-size organization in which close contact is no longer possible. Sometimes the old-line, trusted employees are unable to develop personally or professionally to function well in the larger, more complicated organization. These employees may even blame the contractor, who must try to accommodate and cope with both progress and personal loyalty.

Replacing top management personnel is also a key challenge that many companies are unable to overcome. All too often there is no formal succession plan in place. Most construction companies have a limited number of qualified personnel to choose from during the succession process, and many are reluctant to go outside their own ranks for new leadership, even if the need arises suddenly. This reluctance is even more common in smaller and more closely held companies, where the new leader will at best have worked with and learned under the retiring leader and have no outside or independent managerial experience. Regardless of the circumstances, the new leadership's managerial maturity will be tested immediately upon succession, and that is a critical test of the company's maturity as well. If the new leadership lacks managerial maturity in terms of competence, experience, and confidence, the success of the entire company is at risk.

Leadership transition is difficult no matter the circumstances, but in the closely held family businesses that are common within the construction industry, the succession process is even more complex. The job of operating a family-owned business is often grievously complicated by friction arising from rivalries that exist between the various family members who hold positions in the business or derive income from it. Oftentimes these barriers prove to be insurmountable and can ultimately cause the company to fail.[5]

> *Leadership transition is difficult no matter the circumstances . . .*

7.8 DELEGATION OF AUTHORITY

An important skill needed by contractors who are managing a growing business is the ability to delegate authority. Delegation has two key benefits: First, it builds the company's managerial maturity because the contractor provides their subordinates with experience in making business decisions; and second, it lessens the workload that is placed on the contractor. However, the true delegation of authority within the construction industry is a problem in and of itself. Oftentimes the very personality traits that cause someone to want to become a contractor, such as the desire to be their own boss, makes delegation difficult and in some extreme cases impossible. There are far too many large construction organizations with layers of vice presidents and managers that will not or cannot make even the smallest decisions without the contractor's input. The contractor often thinks it's a good system of checks and balances, but they can't be every place at the same time and end up making every major and minor decision in the company, all while wondering why they're working such long hours. This type of setup is commonplace in the construction industry and is just a mistake waiting to happen. When employees are given titles and not the corresponding authority, they will eventually leave or mentally retire. Few will put the required effort into quality decision making when they know they cannot make a real business decision on their own without clearing it with the boss.

7.9 TEST OF DELEGATION

The test of true delegation of authority is when a person is allowed to make a mistake. That is, when a person to whom authority is delegated can eventually makes unsupervised decisions that may or may not be correct. Contractors frequently resist the suggestion that they should allow their subordinates the latitude to make a mistake because mistakes can cost the company money. Why not, they reason, allow people to make decisions in their area of authority but then require them to check in with the contractor to receive approval prior to implementation? This approach may be a valid training procedure, but trainees do not run construction companies. Eventually decision makers must be given the opportunity to sink or swim on their own abilities and live with the consequences of their own decisions. Of course, mistakes will occur—and some will cost money—but the more important consequence is that personnel will learn from these mistakes and increase their capacity for future success. Mistakes are an inevitable component of management development, but the reward is that developing top managers can genuinely take some of the workload off of the founder or top manager. This also contributes to the company's long-term success by developing the top management capability to lead the company through leadership changes.

The inability to truly delegate has affected the capacity of numerous contractors to hold on to key personnel because lack of delegation drives them away. Personnel who are qualified to lead in the construction business will strive to become contractors, and these types of qualified personnel will not stay long with an organization that is unwilling or unable to delegate true authority. This is even more challenging for relatives in family businesses because those in line for succession oftentimes are not allowed to practice decision making. These members are commonly unable to leave the business due to family reasons and are forced to endure the "no-real-authority" syndrome until their first binding decision is made after they inherit the company. The lack of true delegation of authority sets companies up for failure because managers are unable to truly develop unless they are accountable for making their own decisions. When employees are not given the authority to make their own decisions, they are less likely to develop their own judgment.

Achieving managerial maturity doesn't just happen; the need for it must be recognized and its skills learned or hired. Understanding the importance of real delegation of authority and managerial development to a growing construction company is the first step toward it achievement. Educator and management consultant Peter Drucker writes: "Entrepreneurship is risky mainly because so few so-called entrepreneurs know what they're doing. They lack the methodology. They violate the well-known rules."[6]

> *The inability to truly delegate has affected the capacity of numerous contractors to hold on to key personnel . . .*

> *The true delegation of authority is when a person is allowed to make a mistake.*

7.10   MANAGERIAL MATURITY CASE STUDY

Consider a building construction company worth several hundred million dollars that has had several executive vice presidents in addition to a score of other vice presidents over its 50-year lifespan. The growth of the company was skillfully managed by its founder for years. The founder presided over his huge organization with a strong hand, had the ability to juggle numerous tasks and responsibilities at once, and worked long hours. To the casual observer the founder had a strong management team to back him up and carry part of the load, but upon closer scrutiny there was very little true authority among the senior executives. There was loyalty and plenty of hard work and long hours, but most of the serious decisions were made or given final approval by the founder himself. As he aged, he spent less time at the business, but given the size of the company the important decisions could still be handled by him even if they had to be delayed awhile. Because he maintained all of his personal contacts in the political and business community, the founder remained a critical element in the continued success of the company even after he formally retired.

After the founder reached full retirement, the company was managed by a team composed of some members who had come up through the field construction side of the business and others that were financial executives. There was considerable friction between these two factions. The financial executives who dominated the board of directors, which was made up entirely of insiders, continually attempted to minimize the influence on the business of the field construction people in senior positions. In several instances they attempted to remove senior managers with field experience from the board of directors as they were thought to lack sophistication.

Although the founder was completely retired and in ill health, his influence was strongly felt in the board room of the company even when he did not attend all the meetings. The construction personnel in senior positions had worked for the founder for many years and spoke proudly of the history and values of the company. As a result, they were working hard to continue doing business in a similar manner. The financial personnel on the board, however, pushed multiple changes but were unwilling to fight for these changes with the founder present in the background. The company remained profitable during this period, but the differences among top management grew.

A number of years after retirement the founder passed away. Within months, a board meeting was held, which barely made a quorum because several members were at job sites. At this meeting, a new management committee was formed to govern the day-to-day operations of the company and the chief financial officer was elected president. Within a year the entire construction organization was subordinated to the administrative/financial department, and a number of key construction people quit or were fired.

The company expanded rapidly and more than doubled in size in less than three years following the founder's death. The profit picture was not very good, but management attributed this to the cost of growth and predicted it would improve over time. The introduction of new computer equipment and systems along with the distraction of moving to new corporate headquarters produced an administrative nightmare: Data input was weeks late, and some critical data was lost. This caused a serious delay in producing a certified financial statement at year-end, and when an interim statement six months into the next fiscal year showed operating losses and the prior year's statement wasn't produced yet, the company's surety became concerned and bonding was restricted. Several months later the prior year's results were determined to be a loss in excess of eight percent on volume.

The company operated for six years after the retirement of the founder, three of which followed the founder's passing. The first three years equaled historical performance, and the fourth year was marginal. Because of the time lag from the founder's retirement, the cause of the company's difficulties was not originally associated with the loss of the founder. But what was not recognized was that the founder's influence was felt after his complete retirement and until his death in a way that compelled management to continue to do business with the values, goals, and procedures he had established.

At first glance this might appear like a change in key personnel as the cause of the failure and perhaps even that the founder was in fact a victim. The organization did lose the founder—first to retirement and then death. In addition, they subsequently lost several very key construction people. Looking deeper, consider the reality that every company founder will eventually be lost to the organization and that eventuality should not and does not have to place every enterprise that goes through it at risk. A change in key personnel happened here, but it was predictable and not untimely; therefore, the cause of the problem was the lack of preparation for it. The real cause of this business failure was the refusal or inability of the founder to delegate authority in advance of his retirement and train his replacement.

As a direct result of the founder's inability or refusal to truly delegate authority, there was no one in the organization that had any real authority prior to the founder's retirement. Consequently, no one who had really run their own department let alone an entire construction company. Everyone did their own job as the founder had taught them, and he was regularly consulted if any exceptions were encountered. There was no lack of talent in this company, although more aggressive managers didn't stay long. What the organization lacked was initiative. Everyone followed the founder, but they worked for him, not with him. There weren't any self-starters; they were jump-started by the hard-driving founder.

The success of this company can and should be clearly attributed to the aggressive and talented founder, and its failure is clearly attributable to the lack of management development needed to sustain the company's success once the founder's skill set was removed. It is every manager's responsibility to tutor and prepare his successors, and the founder of a successful company is no exception. In this case, the founder was either unaware of this fact or was unable to successfully address it, which ultimately indicates that he had outstripped his management ability and lacked the managerial maturity to lead the business at the size to which it had grown. Some may argue in this case that building and successfully managing a business of that size proves the founder's management ability. But if the company's continued success relied solely on the founder's continued personal involvement, then it was not truly a viable commercial enterprise. It is critical for business owners to recognize the need to train and prepare their successors if they intend the business to succeed them and to understand that running a business includes the use of and delegation of authority. The cost of mistakes or difficulties associated with delegating real authority to subordinates is a cost of doing business.

7.11   SUMMARY

There are numerous successful construction companies that will be at risk when the founder or current leader is no longer with the firm because of a lack of realistic successor training. They will not all fail. The new leadership may have the innate ability to do the job. However, they are all at risk because they face the transition without benefit of a trained replacement. A trained successor is someone who has been given the authority to practice running the company (or part of it) and has demonstrated their ability and honed their skills on the proving ground of real authority.

Many contractors claim they already have too much to do in managing a growing construction business to be able to deal with time-consuming successor training. This attitude is just another nail in the coffin of growth because if an enterprise cannot control its growth long enough to train people, they are putting the company at unnecessary risk. Other contractors may say, "Let them learn it the way I did." This may be logical in regards to a construction start-up, where mistakes are made on a smaller scale and are more easily recovered from. Additionally, learning is accelerated by the rapid-fire issues that confront the smaller or start-up companies.

However, "learn it the way I did" does not mean taking over a moving train without realistic training. It does not mean taking over a company that the founder has practiced on for 30 years to get it to its current and largest size. And it does not mean taking over a company that even the founder has struggled to manage. Anyone who uses this sink-or-swim training method should not be surprised at its randomly selective results.

CHAPTER REVIEW QUESTIONS

1. From where do most large-size construction companies originate?
   a. Mergers of multiple small- and medium-size construction companies
   b. Smaller start-up organizations that evolve over time
   c. Splintering and restructuring from existing large-size construction companies
   d. Smaller start-up companies that have detailed and sophisticated strategic plans for large-scale growth from the moment they are founded

2. Which of the following are telltale signs that a construction company is trying to manage more than its leadership team is actually capable of (circle all that apply)?
   a. Series of internal gripes or small problems
   b. Complaints from long-time trusted employees
   c. Increased turnover among middle managers and field forces
   d. Increase in owner and designer complaints
   e. All of the above

3. What is the point at which growth will outstrip management ability?
   a. As soon as the company expands to add a new type of work
   b. Once the company has doubled its business
   c. An easily identifiable time because the contractor and management team will "know it when they see it"
   d. A not easily predicted time because the threshold is different for each contractor and management team

4. Which of the following are included among the benefits of delegation of authority (circle all that apply)?
   a. It builds the company's managerial maturity.
   b. It takes away decision-making power from the contractor.
   c. It lessens the workload that is placed on the contractor.
   d. It minimizes the risk of mistakes.

5. What is the most frequent challenge faced by companies that replace top management personnel?
   a. There is no formal succession plan in place.
   b. The existing management team resents the new personnel.
   c. The new personnel are overqualified for the position.
   d. The new personnel lack confidence in existing personnel.

CRITICAL THINKING AND DISCUSSION QUESTIONS

1. What are some of the managerial challenges that accompany rapid company growth?
2. Describe a scenario where a contractor has and has not truly delegated authority to their management teams.
3. Why is it an advantage for an organization to be managed by more than one person?
4. Why is it important for the contractor to delegate authority in such a way that employees are allowed to make mistakes?
5. What are some of the risks that are associated with an organization that lacks a formal succession plan?

NOTES

1. Morgan, D. (1989). "Planning for Profits." *Contractor* (January/February).
2. Flamholtz, E., and Y. Randle (1990). *Growing Pains: How to Make the Transition from an Entrepreneurship to a Professionally Managed Firm.* (2nd ed., p. 17). San Francisco, CA: Jossey-Bass.
3. Lewis, V. L., and N. C. Churchill (1983). "The Five Stages of Small Business Growth." *Harvard Business Review*, 61(3):30–50.
4. Ibid.
5. Levinson, H. (1971). "Conflicts That Plague Family Businesses." *Harvard Business Review*, 49(2):90–98.
6. Drucker, P. F. (1999). *Innovation and Entrepreneurship* (p. 29). New York, NY: HarperCollins.

8

ACCOUNTING SYSTEMS

8.1 ACCOUNTING AND INFORMATION MANAGEMENT

This is not an accounting or information management text and there is no attempt here or in other sections of this book to recommend particular accounting methods or systems. Rather, the subjects of data management and accounting are addressed from a responsibility and risk perspective because many financially distressed contractors attribute their condition to their accountants or accounting systems. Many were shocked to discover the extent of their financial difficulties. The fast pace of the typical construction business and the fact that major projects can last for years makes early detection of financial weakness difficult. This is especially true in growing businesses where growth tends to cover up poor performance.[1]

While most contractors and senior construction executives have a technical background and few have formal accounting education or training, they are not exempt from responsibility for selecting, overseeing, and managing the data management and accounting operations of the organization (Figure 8.1). They obviously are not responsible for completing the accounting work themselves but must use basic logic and investigative reasoning to determine the viability and appropriateness of results that are generated by the accounting systems and accounting personnel. A captain of a battle ship may not know how to operate the engine systems on the ship, but he is certainly responsible to see that they function properly, that the people who operate them know what they are doing, and that the results are appropriate.

8.2 TYPES OF SYSTEMS

There are as many ways to handle the books in the construction business as there are contractors. The methods range from numbers jotted on the back of an envelope to the most sophisticated computer systems. Most large contractors' accounting systems have evolved from when the company was significantly smaller. These accounting systems frequently combine components from the previous system with components from the newer system for current use. Components from a third system may even be included in order to prepare for the future.

Figure 8.1   Managing the Systems And Data Is an Important Function in a Growing Organization Because Knowing Accurately Where You Are at Any Given Point Is Critical to Success.

8.3   WHO IS RESPONSIBLE?

We are not critical of the preceding situation where accounting systems include multiple generational components, nor are we recommending a different way of keeping records. We are firmly in favor of the "whatever-works-for-you" method. But what is of concern, primarily with smaller businesses, is the lack of responsibility that some contractors show toward the accounting aspect of their business. Some even refuse to accept responsibility for the accuracy of any of the separate systems in their accounting and other departments. Yet in truth, an owner/ manager is essentially responsible for the entire operation, not just for those functions they happen to enjoy doing. To abdicate responsibility for accounting ensures that problems will occur. This is not to suggest that contractors or top managers do the accounting work themselves. But they must be adequately informed and remain involved in the function to be confident of its accuracy.

When an outside accountant puts the numbers together in a periodic audit or to prepare financial statements for the company, they are getting the information from the company's internal systems. Outside accountants are not there every day to see that the systems and data are handled properly. Nor do they have the firsthand knowledge of the day-to-day operations that is needed to identify whether items are missing or if the numbers make sense. This is one of the hardest realities for some contractors to understand. Many find it easier to shrug and say, "What do I pay these accountants for?" The primary responsibility for accurate accounting information always remains with the top management of the company.

Accounting problems arise almost universally not from the accounting systems, but rather from the information fed into them. The system is only as accurate as the information on which it operates. The information flow from the operations side of the business to the accounting department must be accurate and timely so the system can compile appropriate data. This is critical to generate the reports necessary to run the business and to produce an income statement. Accounting information is extremely time sensitive because income statements cover a defined period of time, whether one month or a year.[2] Accounting must be a hands-on concern of contractors, large and small alike.

> *The primary responsibility for accurate accounting information always remains with the top management of the company.*

8.4 ACCOUNTS PAYABLE

After accepting responsibility for the accounting functions of the business, top managers need to recognize the potential for problems and errors in data collection and manipulation and try to avoid them. Many managers have difficulty understanding the importance of appropriately managing all data used to compute accounts payable. If an organization doesn't know what it owes, then it can never know its financial condition. Accounts payable are a problem right from when bills are received by the company because invoices aren't necessarily dated and received in the same month the charges were incurred. In fact, it may be easier to understand the problems with payables if, for the purposes of this discussion, we substitute the term "liabilities." Used here, liabilities mean that the company owes something whether or not they have an invoice for it.

For instance, a company incurs a financial liability for material the minute it is received and accepted at a construction site whether or not the company has a purchase order, invoice, or delivery ticket. Further, if the material is delivered on June 30 and the bill is not received until July 15, it is still a June liability. The accounting department may not call it a June payable because they had no invoice in June, payment wasn't due in June, and the material may not even have been used in June. The accounting department may have no way of knowing that the company had incurred the liability for payment of the material in question in June. In fact, almost everything delivered to the job sites in June doesn't get billed until July. Consequently, the organization that records accounts payable as the date on the invoice (as opposed to delivery date) or as the date the invoice is received will always have account payable information that is up to one month behind.

That is the problem. Accounts payable in construction are sometimes out of sync with actual liabilities incurred. Consequently, construction enterprises often owe more than their accounts payable reveal at any given point in time. Discrepancy is acceptable as long as it is small. We can contrast this with the reality that every construction organization that prepares a pay requisition from an owner on July 3, 5, or 7 for work performed in June records the accounts receivable as a June event in June sales and rightfully so. But taking these events together demonstrates the position taken here that accounts payable and accounts receivable are sometimes out of sync in a smaller construction business.

8.5   DISPUTED INVOICES

The management of data concerning liabilities and the accurate recording of accounts payable is a prerequisite to accurate financial information. Disputed invoices are common in any business whereby a supplier or subcontractor invoices the company for an amount that the company disagrees with. Many organizations have excellent systems to assure that disputed invoices do not get paid, but the same systems sometimes prevent the liability form being recorded. The proper way to record accounts payable are in the month they were incurred and not when they are dated or received. This includes disputed payables. We have observed $100,000 invoices left off accounts payable because they were returned to the sender with a $1,000 disputed amount with neither the disputed $1,000 nor the undisputed $99,000 recorded as a payable. Some companies inadvertently do this regularly, which can result in substantial distortions in accounts payable.

To make matters worse, when disputed invoices are finally recorded, they are often recorded as of the month they are finally approved because the books for previous months have been "closed"—the accounting term for finalized and reported. Some disputed invoices will inevitably span the close of an accounting period, causing distortion. Some may wonder if these amounts really matter on the overall scale of things in a construction operation. Account payable distortions of this nature, if confused by field personnel to believe that certain completed work costs less than reality, can lead to inappropriate estimates of, or management of, remaining work items. If several of these occur at the same time, the distortion can be substantial. The point is that a company cannot follow procedures that confuse the system and deliver distorted results. There is enough potential for error in data collection without building it into the system.

> *There is enough potential for error in data collection without building it into the system.*

8.6   CASE STUDY

A smaller contractor located in the northern part of the country was having serious financial difficulty, but his financial statement showed profits for the 25-year history of the company. The contractor was baffled because cash had run out and debt had increased to his credit limit, all while his financial reports showed steady profits. His annual volume was about $10 million, done mostly between April and October. His fiscal year ended in October, which was usually a high-volume month because at that time of year the organization was always pushing to finish as much work as possible before the cold weather. For the prior several years, September and October had been the biggest profit months and November and December the worst. No one noticed that a lot of September and October accounts payable never hit the books until November and December. However, this was not the only distortion affecting the company.

A tour through the payable approval process proved enlightening. It began in the contractor's office where the incoming mail was put each day. The contractor would look over all the invoices as his way of keeping up on them and would initial which project manager or superintendent they were to go to for approval. Then he would put them in his out-basket. There were a number of invoices there at the time observed. and the amounts were recorded.

Each day a clerk would distribute the invoices from the contractor's out-basket to the in-baskets of the appropriate project managers or superintendents who came in at the end of each week, and they would approve or disapprove the invoices along with their other paperwork. The several piles of invoices in both the in- and out-baskets of the project managers and superintendents were recorded. From the received date stamp it was easy to determine that the length of approval time on these invoices was generally one or two weeks because the project managers and superintendents usually took the invoices back to their job sites for an accurate check of quantities and returned them the next time they were in the office.

Each Thursday these approved invoices were moved to the controller's in-basket where he looked them over as his way of keeping up. This was always done on Thursday because the bookkeeper set aside Thursday to code all approved invoices with the vendor code, which would then be input into the computer by the part-time person who came in once a week on Thursdays. What management didn't know was that to avoid being rushed, the data put into the computer was that which was coded the week before. This meant that weeks of delays were built into the data collection system. If there were quantity disputes or if a project manager or superintendent got behind in their paperwork, a month delay in approving invoices was not uncommon. When all of the invoices that were in the various in- and out-boxes were totaled, the result was over $400,000 in unrecorded liabilities (payables)—most of them over 30 days old and some over 60.

Because of the continuous time lag, this company never had an accurate picture of what it really owed or what its true cash needs were, let alone its profits, if any. In addition, the business was seasonal and the fiscal year ended at the end of the season, and the time lag on account payable approval was during a period when unrecorded liabilities were at their peak and sales was also at peak. The approval delay built into their system gave the organization a completely false picture. When this was pointed out, there was great debate from the accounting people because they had what they believed to be an accurate computer-generated payable aging report that they put great faith in. The aging hadn't appeared to be all that bad.

Further confusing the account payable system, the invoices were entered by the part-time data processor using the date on which they were entered and not the date of the actual invoice. Therefore, accounts payable were being aged against an erroneous date. The built-in delays in this case crept into the system over time, and in this company were extreme. However this type of information mismanagement has been observed in numerous construction operations of various sizes, and it should be noted that sizable organizations will experience greater distortion due to the sheer magnitude of liabilities they deal with.

8.7 RECORDING LIABILITIES

The appropriate way to record invoices as accounts payable is upon receipt. Disputed invoices can be recorded as "not approved for payment" but must be captured by the system as potential liabilities before they are passed around the company, and the accounting department loses track and control of the information. It is inappropriate to work under the assumption that invoices are not liabilities until approved. If a disputed invoice is disapproved, or substantially reduced at a later date, it is prudent and conservative to use a reversing entry rather than not record every disputed invoice. It is a matter of control and accuracy.

For the company to know where it stands at any given point, it needs to know what it owes, and the invoices, as the source documents, must be captured by the accounting systems at the earliest possible time—when they arrive in at the company (Figure 8.2). Passing the invoices around the organization

© iStockphoto.com/shironosov

Figure 8.2   Information Must Be Put into the Systems in a Timely Manner for Them to Produce Accurate Results Quickly Enough to Be Useful.

before recording them can allow so much time to elapse before capturing the data as to lose control of accounts payable altogether. The problem is one of knowing the company's real liabilities at any given point in time so management can tell if the organization is making or losing money and how much. If profits are overstated, financial decisions may be made to use resources that do not exist.[3]

8.8 ACCOUNTS RECEIVABLE

Delayed recording does not typically afflict account receivable systems of construction companies. Most organizations record that which is owed to them in a timely manner. It appears that just about all contractors record sales correctly in the month the work was performed irrespective of when the requisition or invoice to the customer was prepared or dated. There is, however, a problem observed with the aging of receivables, which in some cases can cause distortions in the profit picture.

When receivables appear uncollectable, they should to be written off in a timely manner to avoid distortion. That is, they should be taken off the books as assets and, of course, entered as bad-debt expenses. Consequently, the bottom line will be affected. In difficult times, construction organizations are reluctant to write down old receivables considered to be uncorrectable or old retainages that have for practical purposes been waived. The reluctance is based on the negative impact on financial statements. If a company is doing this—manipulating receivables—to maintain their credit line, they need to at least make the adjustment in their head and know their real position as opposed to the position on the financial statements. It is imperative that managers know and react to their real financial condition.

It's not good practice, but if for any reason the financial statements don't reflect all of what is known to be the case, then management must adjust the numbers in their head to reality and manage the business using the correct figures. It's best to take write-offs in a timely fashion. Only management knows what is timely, and they can't expect their accounting department to know when to write something off. The systems used by most accounting departments keep every receivable on the books until it's received. Accounts receivable need to be regularly reviewed, preferably monthly, by responsible senior responsible managers and adjusted for accuracy.

8.9 TIMELY DATA ENTRY

Accounting systems and subsystems are primarily methods and procedures for capturing information about operations in order to provide management a numerical picture of how the company is performing. The systems do little in and of themselves. They don't produce a product. They just collect data, fold it over in several ways, and give the same data back to management in a usable form. The systems can't improve the data, but they can provide an altogether false picture if someone misuses or confuses them. Entering information out of sequence will generate a false picture.

If the information is only a little wrong, then it's only a little useless. If it's a lot wrong, it's not only totally useless but is downright dangerous. With this in mind, it is suggested again that every contractor, particularly smaller businesses, needs to understand the accounting systems his or her company uses. They need to be close enough to the output data to verify monthly that it makes sense as it relates to the prior month's operations. Data collecting is pretty straightforward: Use a system that's right for the organization and make sure everyone inputs the information correctly. Most companies start out with appropriate data collection methods, but over time someone changes a procedure here and alters a process there and slippage occurs. Time delays get built in without the users of the systems knowing it. Management must be diligent in monitoring data collection processes regularly for appropriateness and timeliness. A good safeguard is to review every detail of each system at least annually.

> *If the information is only a little wrong, then it's only a little useless. If it's a lot wrong, it's not only totally useless but is downright dangerous.*

8.10 SUMMARY

Companies operating without good data collection and accounting systems are at risk because a construction operation cannot be properly managed without accurate and timely accounting information. Similarly, if companies have the systems but

> *Modest profits in the construction industry provide for no margin of error in accounting for them.*

do not receive and use the reports on at least a monthly basis, they are not managing safely. Management needs to know enough about the systems to feel comfortable in judging that the output numbers make sense in the context of current operations. The collection, compilation, and calculation of construction accounting data are so complex that the people who do the work begin to revere the results. That is, they begin to respect the resultant numbers too often without checking or in some cases not even knowing if they make sense or fit with current trends or expectations.

The contractor who knows enough about how the data is collected, where the numbers come from, and how the systems work can and should question results that don't fit normal patterns or expected results. Modest profits in the construction industry provide for no margin of error in accounting for them. To manage their financial affairs properly, contractors must regularly analyze their historic financial data.[4]

CHAPTER REVIEW QUESTIONS

1. Which of the following are key challenges with using accountants from outside the construction company?
   a. They do not have firsthand knowledge of the day-to-day operations.
   b. They are not there every day to see that the systems and data are handled properly.
   c. They are easily able to access the construction company's internal systems.
   d. a and b

2. At what point in time does a company incur a financial liability for material?
   a. As soon as the need for the material is identified
   b. As soon as the material is delivered and accepted at a construction site
   c. As soon as the company receives a purchase order, invoice, or delivery ticket
   d. As soon as the material is put into place in its final position within the construction project

3. A contractor should understand which of the following about the company's accounting systems?
   a. How the data is collected
   b. How the systems work
   c. The timeliness of data entry
   d. All of the above

4. How often should accounts receivable be reviewed for accuracy by responsible senior managers?
   a. Weekly
   b. Monthly
   c. Quarterly
   d. Annually

5. Large distortion of year-end results is most common in what kind of company?
   a. A company that is not growing
   b. A company that is slow growing
   c. A company that is fast growing
   d. None of the above—it is a universal problem among construction companies

CRITICAL THINKING AND DISCUSSION QUESTIONS

1. Why must top management be adequately informed and remain involved in the accounting function to be?

2. When does a company incur a financial liability for material and why?

3. How would you treat disputed invoices from your vendors in a construction enterprise?

4. What is the difference, if any, between liabilities and accounts payable?

5. How does the timeliness of entering accounting data impact a construction organization?

NOTES

1. Hillebrandt, P. M. (1985). *Economic Theory and the Construction Industry* (pp. 12–19). London: Macmillan.

2. Block, S. B., and G. A. Hirt (1989). *Foundations of Financial Management* (p. 26). Homewood, IL: Irwin.

3. Schleifer, T. C. (1993). Seminar presentation: "Managing Risk Through Business Planning." Associated General Contractors of America Annual Convention: March 20, 1993.

4. Bersch, D. W. (1973). "Do You Manage or Just Watch?" *Constructor,* (September).

9

EVALUATING CONTRACT PROFITABILITY

9.1 MEASURING PERFORMANCE

Accounting, the language of business, speaks to company owners and managers through numerous reports and, at least annually, through financial statements that include a balance sheet, income statement, and statement of cash flows, usually prepared by independent outside auditors.[1] The methods and procedures used in the preparation of financial statements for construction enterprises in the United States are regulated by Generally Accepted Accounting Principles (GAAP) rules.[2] However, there is some flexibility in the reporting of transaction within the accounting rules. The same event can result in a different measurement of income[3] such as with contract profitability on uncompleted projects. Because all contracts do not conveniently start and finish within a construction company's fiscal year, the measurement of profits is complicated with much room for error or manipulation.

A construction company handles a lot of money that isn't actually theirs. The company gets to keep only the profit, which most agree is too small for the risk (though no one can convince construction industry employees or the general public that it's small at all). Most medium-size and larger contracting firms would be happy to keep 2 or 3 percent of sales if they could do it. The basic problem most contractors have with evaluating contract profitability is that they leave its calculation to the bookkeepers and accountants. That's not to say that bookkeepers and accountants can't do accounting work, but rather that too many contractors and senior managers avoid accounting like the plague. They don't like it and often distance themselves from it, but the calculation of contract profitability, particularly work in progress, requires accurate input of construction information known only to the people actually engaged in the work in the field.

> *A construction company handles a lot of money that isn't actually theirs.*

Contractors and senior managers who come up through the ranks or from construction or engineering schools may never have had even a basic

© iStockphoto.com/tiridifilm

Figure 9.1   The Systems And Methods to Collect And Use Data Should Be Matched Carefully to the Company's Current And Future Needs.

accounting course. They can do math and geometry, but they don't want to deal with debits and credits. Yet when a contractor doesn't participate in the accounting function of the business or has a limited interest, the book-keepers and accountants are left to collect work-in-progress information and develop reports and statements without the help of operations personnel who can say from everyday knowledge whether the information makes sense (Figure 9.1). Some smaller contractors have so little use for the accounting side of their business that they even resist passing along all of the real data to their employees.

9.2   ACCOUNTING FOR PROFIT

The primary reason for being in business is making a profit. Even if a contractor is in the construction business for other reasons, he or she won't be able to stay long without making a profit or at least breaking even. The difference between a profit and loss on some jobs is extremely small and could involve only a couple of good or bad breaks. However, if making a profit on a job is difficult, the only thing harder is accounting for it. Capturing all of the data accurately and in a timely fashion and using it properly to track interim profit or loss and then final profit or loss is one of the most difficult and misunderstood processes in the construction business.

Profit equals sale price minus costs. That's it. And if the job lasts about a week and everyone bills properly and on time, and if you don't forget the fringe benefits or cost of company-owned equipment, and if the company eventually collects retainage and remembers to back in the call backs and guarantee work, then it's easy. For construction companies whose work lasts a little longer, however, it starts to get complicated. It begins to involve all of the various accounting systems in the organization to determine if the organization is making money or losing it.

9.3 SELECTION OF SYSTEMS

If management is going to account for all of the money that passes through their hands (if only to know how much they get to keep), they need to participate directly in the selection and use of the accounting and bookkeeping systems and data collection methods. The systems must make sense to non-accounting managers and fill the organization's needs as a tool to run the business. The results must be accurate, and it is management's responsibility to make sure they are accurate. It is critical for the contractor himself or, in the case of larger companies, top management to participate in the accounting process. They need not do the actual work, but must check on the systems if the numbers in the broad categories and subcategories are to make sense from month to month. Management can't just come in once a year and do this. They must remain involved in the accounting function each month to make sure it is serving the organization's needs.

When management does not regularly scrutinize accounting results and simply accepts them at face value, inaccuracies go unnoticed. We have experienced overhead costs reported and accepted that were 50 percent of the previous month's, and no one sensed that there must be an error. And we have seen sales reported improperly, with no one realizing that billing did not go out on one of the largest projects the prior month.

A contractor who watches the numbers side of the business continuously knows the approximate month's volume and costs before the accounting department does, and rightfully so. The bookkeepers and accountants can only add up that which is given to them in the normal course of business and put these figures into the proper categories to develop reports. But if vendors don't bill for their subcontracting work for last month, only the contractor or field management (not the bookkeeper) knows the work was done and the cost or liability was incurred.

Some contractors contend that as their business gets larger, they can no longer pay attention to the numbers each month. Yet to manage properly, they need to know on a regular basis whether the information they are getting makes logical sense. Are they certain that all the costs incurred are included in the report? Does the report show a 20 percent gross profit when they know that's just not the case? Making sure the reports make sense may mean acquiring better systems and in smaller company a controller instead of a bookkeeper, or it may involve hiring a qualified chief financial officer instead of a controller (depending on the size and stage of growth of the company). If the contractor genuinely does not have the time, training, or ability to manage the accounting function, then a board of directors or advisors should be considered as a safety valve. Another option is to involve outside board members, who always seem to know if the numbers make sense because they are usually far enough away from the trees to see the forest.

> *Some contractors contend that as their business gets larger, they can no longer pay attention to the numbers each month.*

9.4 PERCENTAGE OF COMPLETION

The state of the art in construction accounting has improved over recent years but continues to be an area of great concern. The primary problem lies in the fact that an independent accountant has no way to verify the accuracy of the figures for percentage of completion that the contractor provides for each project (Figure 9.2). For that matter it is even difficult for contractors themselves to determine if a project is 40 percent complete or 38 percent complete. You can't simply stand on a site and say, "This project is 62.5 percent complete." The exposure here should be obvious in an industry where the real profits for any

© iStockphoto.com/Mikosch

Figure 9.2 Determining the Percentage Complete with Any Accuracy of an Ongoing, Complicated Project Is Difficult at Best.

sizable contractor are in the low single digits. An error of 1, 2, or 3 percent in the estimated percentage of work completion at statement time can make the rest of the accurate information on a statement erroneous.

There would be some relief if the different estimates for percentage of completion for all projects would average themselves out, with some being high and some low. Yet since the people making the estimates usually have their performance and bonuses on the line, it is perfectly human for them to estimate the percentage of completion on the high side. For a construction company of any size, few projects start and end conveniently in any one fiscal year. This means that a large part of the income claimed for a year will be based on the estimated percentage of completion on all contracts except those that were closed out during the year.

A construction company therefore records a portion of the estimated profit on work in progress as earned income. If they guess the amount of work completed to be higher than it actually is (by underestimating the cost to complete), they are claiming more profit than earned in the period. If a company is growing, it will have more and more work in progress each year, and if the estimates of completion percentage are slightly on the high side, the previous errors in completion estimates will be covered by new projects. That is to say, if the percentage of completion of projects is exaggerated in one year, it will probably go unnoticed as long as there is more work in progress the following year.

Some might claim that eventually it all comes out in the wash because when the projects are closed out, the totals are reported. This may be true, but in the construction business management needs to know accurately where the organization is in real time—if they are making a profit or loss on the present work and how much. One of the greatest exposures a construction organization faces is to continue doing the work in a way that doesn't make a profit while believing it is because the accounting reports say so. The danger is that no steps are taken to do the work differently, and all go merrily on their way.

9.5 ESTIMATED PROFIT

There's an even more insidious problem with this procedure if the real profitability of each project is not reviewed periodically. It is necessary to periodically reevaluate the actual contract profit against the originally estimated profit, particularly at statement time. So management needs to determine not only what percentage of the profit is earned on uncompleted projects, but also what the current projected profit is.

> *Management can't do much about losing projects if they don't know the losses are occurring until the jobs are finished.*

9.6 CASE STUDY

This particular construction company had been in business nine years, and business had grown to a volume of $70 million in that time. With such rapid growth, each year there were more and larger projects in progress and, of course, in progress at statement time. For each year including the last, this company's certified financial statement indicated strong profits and, after nine years, a very substantial net worth.

Reality was that the company had a negative net worth, and the negative worth did not simply result from recent losses or losses on current projects. As the company's jobs became larger, they also ran longer. Consequently, there were not only more new projects each year, but also more ongoing projects from previous periods at each statement date. The total profits from the increasing amount of work, based on the percentage-of-completion method, were substantial and always much greater than the previous year. This meant that a losing job reported as profitable in a previous period was well covered, and the total profit for all jobs for the current period was a positive and substantial amount.

What had been happening, of course, was that profits were claimed each year for jobs that were reported as a loss in subsequent years because they never actually made a profit. There were quite a number of these, but with so much growth, the statements always looked good and revealed company-wide profits. Finally, the cash ran out and a subsequent study of all past and existing projects showed some startling results. The overall picture was a disaster, but what was most interesting was that of all completed projects for the past nine years, more than 65 percent of them had lost money. These losses were covered by the overstated percentage-of-completion estimated profits of the escalating amount of work in progress. During this time the ongoing jobs were shown to be earning their originally estimated profits, but many would eventually be reported at a loss. Management didn't do anything to correct the losing projects because they didn't know the losses occurred until the jobs were finished.

9.7 PERCENTAGE OF COMPLETION METHOD OF ACCOUNTING

To more thoroughly understand how the previous case could happen in spite of a large internal accounting staff and certified annual financial statements, a closer look at the percentage of completion method of accounting is warranted. The accounting methods and procedures used in the construction industry are, for the most part, the same as those used in the manufacturing industry. In both industries, work that is in progress at the close of an accounting period must be dealt with and calculated.

GAAP rules provide for the use of the percentage of completion method, which allows a company to record in an accounting period profit on the completed portion of work in progress. For instance, if work on a project is 40 percent completed at the close of an accounting period, then 40 percent of the total anticipated profit on the project can be recorded as earned in the period[4]—even if the work in question has not been invoiced or sold. In construction, work in progress can be of a nature that the total cost of completing the work may not be known at the 40 percent completion point so the total anticipated profit is unknown (Figure 9.3). An estimate of the total anticipated cost or an estimate

Figure 9.3 It Is Difficult to Even Pin Down the Percent Completed of a Partially Completed Project

of the cost to complete the work is required by GAAP rules.[5] The contractor, not the independent accountant, is responsible for the estimate of the cost to complete the work. The auditors must review the numbers but are not responsible for their accuracy under GAAP rules.[6] The independent auditor can attempt to test the validity of the estimate by checking it against the cost of recently completed work, but construction estimating skills are required to do this.

The significance of the treatment of work in progress in the construction industry and its potential impact on financial information reported may be better understood by contrasting it with the treatment of work in progress in the manufacturing industry. Work in progress for most manufacturing operations represents a small portion of total revenue for a year or accounting period because production cycles are generally short. The automobile industry boasts of producing vehicles in less than a day. Even custom-made heavy machinery seldom has a production cycle of more than months. Therefore, the manufacturing enterprise will have only a small portion of total revenue represented by work in progress at the close of an accounting period, which means that the accuracy of the total revenue amount is not materially affected. The potential for error in accounting for work in progress in construction is greater because revenue from work in progress is based on an estimate.[7]

9.8 CONSTRUCTION—WORK IN PROGRESS METHOD

The construction enterprise differs from manufacturing in that the project production schedule is extremely long. Many construction projects last longer than one year, and those that last less than a year will often span the close of the fiscal or accounting year. It is common for a construction company to have 50 percent or more of total revenue in any given fiscal year represented by work in progress, and for large companies it could be more. Construction projects, typically one-of-a-kind, are extremely difficult to estimate initially and even more so once they are in progress. If a large percentage of total revenue for a construction enterprise is represented by work in progress in any given fiscal year and the anticipated cost of completion is difficult to estimate, total revenue can be in error, which would produce false profit figures.

A work in progress schedule is part of the supplemental information in most construction company financial statements. These schedules, sometimes called gross profit schedules or contract status reports, are the source document for the first entry on the income statement, namely, revenue.[8] Revenue is also called sales. A typical work in progress schedule is shown in Table 9.1 and the inputs and outputs are diagrammed in Figure 9.4. The headings may be placed differently in some firms but the input data and formulas used to calculate the other entries are commonly used.

The four columns noted "input" represent the only data that is required for calculation. All other columns are calculated from the input data. The first entry (column 1) is "contract price," which includes signed change orders and can be audited through source documents.

The next entry, "amount billed to date" (column 8), can also be audited through the contractor's internal documents. However, the most recent invoices may not

Table 9.1   Work in Progress Schedule

CONTRACT DESCRIPTION	Column 1 INPUT CONTRACT PRICE	Column 2 (4+11) ESTIMATED TOTAL COST	Column 3 (1–2) ESTIMATED TOTAL PROFIT	Column 4 INPUT DIRECT COST TO DATE	Column 5 (4/2) PERCENT (%) COMPLETE
JOB A—PROFIT (UNDERBILLED)	$1,000,000	$ 800,000	$200,000	$400,000	50.0%
JOB A—NO PROFIT (OVERBILLED)	$1,000,000	$1,000,000	$0	$400,000	40.0%
JOB A—LOSS (OVERBILLED)	$1,000,000	$1,200,000	($200,000)	$400,000	33.3%

INPUTS AND OUTPUTS FOR WORK IN PROGRESS METHOD

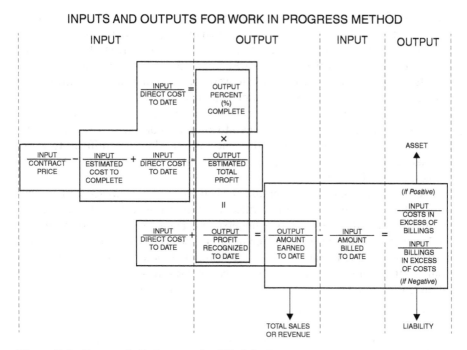

Figure 9.4 Inputs And Outputs for Work in Progress

be approved by the owner and are sometimes reduced at a later date, allowing for potential overstatement of this entry.[9] The third input entry, "direct cost to date" (column 4), represents those costs incurred and recorded through the company's cost accounting system. Timeliness and diligence are required to record accurately all costs incurred and invoiced by vendors. Weakness in the collection and recording of construction costs (accounts payable) can cause entries to be understated.

The entry that causes the most problems in construction accounting, "estimated cost to complete" (column 11), is produced internally by the construction firm and

Column 6 (5×3)	Column 7 (4+6)	Column 8 INPUT	Column 9 (8-7)	Column 10 (8-7)	Column 11 INPUT
PROFIT RECOGNIZED TO DATE	**AMOUNT EARNED TO DATE**	**AMOUNT BILLED TO DATE**	**BILLINGS IN EXCESS OF COSTS**	**COSTS IN EXCESS OF BILLINGS**	**ESTIMATED COST TO COMPLETE**
$100,000	$500,000	$400,000		$100,000	$400,000
$0	$400,000	$400,000	$0		$600,000
($66,667)	$333,333	$400,000	$66,667		$800,000

is difficult for an independent accountant to audit. An accurate determination of the amount of work remaining to be completed at any point in time on an ongoing project can be as difficult as estimating the cost of the work from the outset. The costs incurred on the work completed are commonly used in estimating the cost of completing the remainder of the work; however, it may not be representative of the cost of the remaining work. For example, the cost of brick completed on the first floor of a high-rise building may be less than the cost incurred on the upper floors.

Independent accountants are required to audit the contractor's estimate under GAAP rules.[10] Some indicate they do review it but admit it is difficult to verify the exact stage of the work at a point in time well prior to the time that they are doing the audit function.[11]

The accuracy problem is compounded when the entry, "estimated cost to complete," which is not independently verified, is combined with "direct cost to date," an entry that has the potential to be understated. These two are added together to produce the "estimated total cost" (column 2). The accuracy of profits calculated from this data is questionable because the "estimated total cost" is then subtracted from "contract price" to produce "estimated total profit" (column 3). Refer to the callout calculation using columns 4, 11, and 2 from Table 9.1.

Column 4 Input	+	Column 11 Input	=	Column 2 (Col 4 + Col 11)
Direct Cost To Date		Estimated Cost To Date		Estimated Total Cost

The "percent complete" entry (column 5) is calculated by dividing the "direct cost to date" by the "direct cost to date" plus the "estimated cost to complete." This is called the percentage of completion method, but it might better be described as the "estimated cost to complete method." Refer to the callout calculation using columns 4, 2, and 5 from Table 9.1.

Column 4 Input	/	Column 2 (Col 4 + Col 11)	=	Column 5 (Col 4 / Col 2)
Direct Cost To Date		Estimated Total Cost		Percent % Complete

The percentage of completion method allows "estimated" profit to be recognized in a current accounting period. Yet when the profit figure is put into the income statement, the word "estimate" is dropped even though both entries used to calculate it were products of estimates. "Estimated total profit" times "percent complete" is the calculation used to determine "profit recognized to date." Refer to the callout equation using columns 3, 5, and 6 from Table 9.1.

Column 3 (Col 1 – Col 2)	×	Column 5 (Col 4 / Col 2)	=	Column 6 (Col 5 × Col 3)
Estimated Total Profit		Percent % Complete		Profit Recognized To Date

"Profit recognized to date" is then added to "direct cost to date" to produce the entry "earned to date." This, of course, is the basic purpose of the calculation: to determine the amount of the work in progress that can be claimed as earned in the current accounting period. Refer to the callout equation using columns 6, 4, and 7 from Table 9.1.

Column 6 (Col 5 × Col 3)	+	Column 4 Input	=	Column 7 (Col 4 + Col 6)
Profit Recognized To Date		Direct Cost To Date		Amount Earned To Date

Amount earned to date for each project becomes revenue or total sales and is transferred to the income statement. Revenue for a construction company is a calculation that relies *completely* on the accuracy of the contractor's estimate of the cost to complete their work in progress.[12] If one or more projects do not perform as well as expected, revenue will have been claimed that never existed and will then have to be "paid back" in a subsequent accounting period.

9.9   OVER- AND UNDERBILLING

Over- and underbilling is unique to the construction industry and is a product of the work in progress schedule. As demonstrated earlier, the amount a company has billed for work completed on a project during an accounting period is not taken into consideration in the calculation of revenue or total sales for that period. The entries "overbilling" and "underbilling" are used to "balance" or correct the effect of not using the amount billed. They, in effect, correct the amount that should have been billed according to the work in progress calculations. An explanation of the remainder of the work in progress schedule explains how this is done.

The amount "earned to date" has nothing to do with amount "billed to date" because the latter has not been used in the calculation of the former. "Billed to date" is introduced into the calculation and simply balanced against "earned to date." If "earned to date" is less than "billed to date," it is called "billings in excess of costs", which appears in the balance sheet as a liability. If "earned to date" exceeds "billed to date," it is recorded as "cost in excess of billings" and appears in the balance sheet as an asset.[13]

Reliance on an estimate to produce the total revenue entry on the income statement is a weakness in construction accounting. Applying "costs in excess of billings," derived from the same estimate, as a current asset in the balance sheet doubles the problem if the estimate is understated. In effect, a future loss can be recorded as a current asset.[14]

Table 9.1 shows the effect of changing only the estimated cost to complete on a hypothetical project. Percent complete on the work in progress schedule does not refer to physical completion of the project but only relates the costs incurred to date and the assumptions about the costs yet to be incurred to complete the work. If three different people estimate job A and disagree, the percentage of completion changes dramatically. This example is extreme in that it runs from a 20 percent profit to a 20 percent loss to demonstrate the point: If the estimated cost to complete is \$400,000, the project is 50 percent complete; if the estimated cost to complete is \$600,000, the project is 40 percent complete; and if the estimated cost to complete is \$800,000, the project is 33 percent complete. Therefore, the accuracy of the estimate of the cost to complete the work in progress has a material effect on both the income statement and balance sheet. Table 9.1 demonstrates these dramatic shifts resulting from a change in the estimate of the cost to complete. It should be clear that if the estimate is wrong, deliberate, or accidental, the revenue and profit reported on the financial statements are wrong.

9.10   IMPACT OF TOTAL REVENUE

The accuracy of the total revenue figure is critical because of its impact on the financial statements of the enterprise. Total revenue is the first line in the income statement and directly impacts stated profit or loss. Profit or loss for the fiscal year directly affects assets or liabilities on the balance sheet. The information flow within the management structure is shown in Figure 9.5.

In Figure 9.5 it can be seen that the percent completion of a project is seen most clearly in the field, but a project manager, who is in the office, usually prepares the estimate of the cost to complete. The accounting department then uses costs incurred (and recorded) and the estimate of the cost to complete to calculate the percentage completion of the project rather than field personnel doing it. The accounting department uses this calculation to determine total sales and adjusts or justifies any disparity between the stage of completion of the project and the percent billed on the project.

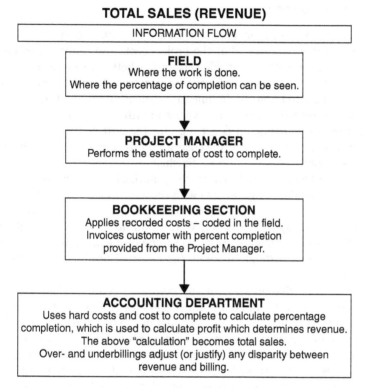

Figure 9.5 The Information Flow within the Management Structure

Because of the effect on cash flow, a prudent businessperson would be expected to avoid underbilling at all times. However, the "cost in excess of billings" entry appears in virtually all U.S. construction company financial statements. The entry suggests that costs have been incurred that have not been billed for. They will be billed for in the future. With the obvious advantage of billing for all work performed in a timely manner and the cash flow pressures of financing the work, it is unlikely that the industry is in a net underbilled condition. Underbilling implies that the people billing the work are not communicating with those building the work or that the estimate of the cost to complete the work is optimistic or in error. The amount billed on a project is not taken into consideration in the preparation of the estimate of the cost to complete the work on the project.

It is apparent from the preceding examples that the percentage of completion method used in construction has the potential for errors, can lead to a false sense of security, and needs to very carefully managed.

> *… the percentage of completion method used in construction has…serious…potential for errors…*

9.11 COST CONTROL

A construction organization must have an appropriate cost control or job-costing system that captures all of the project costs and captures them on time so that it is known, with reasonable certainty, whether or not a job is making money. And the timeliness of information is extremely important because the organization needs to know this in time to do something about it. If the accounts payable system records costs in the wrong month, as described in Chapter 8, then the same misinformation is getting into the cost control system. If sales are recorded timely and some of the costs a month later, then each job will look great on paper until a month after it's finished.

There are numerous sophisticated computerized cost control systems that contain poor input data or good data with the wrong dates. When that happens, these systems become inherently dangerous. Getting the wrong information and relying on it to manage a business is worse than no information at all. This is because almost every potential error and timing problem tends to lose or ignore costs. When the lost costs are ignored, the reported profits increase. But these profits aren't real.

There are many different kinds of cost control systems, and this text does not include recommendations or go into the mechanics of them. Rather, what is discussed is what the system should do and what has to be put into them and when.

A construction organization needs timely profit and loss information by project and by separate activity. A cost control system should collect costs in categories that can be tracked against the originally estimated cost for each activity. A cost control system actually controls nothing, but it does provide necessary information about how the project is running against planned or budgeted amounts, which enables management to better control the project.

9.12 TIMELINESS

The key to the success of a cost control system is getting the information in time to do something about it on the job. The information, of course, needs to be accurate to be of any value and trading off accuracy for timeliness doesn't work. The organization must have both. Therefore, all available information must be input immediately. If a bill for concrete is received, it should not be delayed by sending it to the field to verify quantities and then to accounting to check the extension of the figures. It should be put into the system immediately as an incurred cost. If it has to be changed later, it can be changed in the system by a correcting entry. Having it entered is more important than finding out later there is 2 percent discount or even that the invoice is in error by 50 percent.

There are usually plenty of invoices missing from any system because they haven't been received yet. The fact that the company hasn't received an invoice doesn't mean they haven't incurred a cost. It is preferable to have a system that will understate the profit rather than overstate it because there are enough pressures within an organization and most systems to overstate performance and profits. A company needs such a system for a very simple reason. When profit is overstated, no action is taken. Management is pleased with the way things are going. But when it's understated, they will be less pleased and take steps to improve performance.

> *The key to the success of a cost control system is getting the information in time to do something about it on the job.*

9.13 COST CONTROL VERSUS GENERAL LEDGER

The following is a controversial position and many highly qualified accounting professionals disagree with it. It is widely believed that cost control systems should be integrated with a company's general ledger accounting as all the necessary data is collected by the primary accounting systems. However, the general ledger must be maintained in accordance with GAAP rules as well as to great accuracy, and in many companies the result is that accurate information is not available until far too late in the month to be of much value in managing field operations. If this is the case, other steps can be taken.

The project cost control system doesn't need to tie into the other accounting systems in the company. It's fine if it does, but not if it is a slave to the other system; that is, the system has to wait for data until the other systems receive approvals for accuracy. If the latter is the case, use an independent cost control system that recognizes committed costs (explained later) and don't compare the numbers at all. It is far more important to get this critical management information on a timely basis. The other accounting information can be used for other purposes.

9.14 TRACKING COSTS

The process of accounting for all costs may be more complicated than it has to be if certain costs are tracked unnecessarily—for instance, subcontractor costs. An example would be a building construction project where 80 percent of the project is subcontracted. In this case the costs that can actually be controlled by field management are the remaining 20 percent, and those costs should be tracked closely (Figure 9.6). Put another way, the subcontracted costs are committed early in the process as in a subcontract for a fixed sum: The money can be considered reserved if not spent at the time the subcontract is signed. Excluding change orders, the full amount of this subcontract will be incurred as the job gets done. Knowing how much to pay against the contract as the job progresses is a totally different subject. These figures don't come from the cost control system. It is quite the opposite: When the subcontractor is paid, that information is fed back into the control system. Therefore, it makes good sense to list, in an independent cost control system (one not tied into the general ledger), the full subcontract amounts as committed or spent as soon as the subcontracts are

Figure 9.6 Tracking Costs of Everything in a Complicated Project Is Difficult at Best

signed. Of course, the amount is listed right next to the estimated amount. It is not necessary to wait until the end of a job to know that if the bid was $90,000 for electric and it is awarded it for $85,000, the company made $5,000 on the line item (or lost money if it was awarded for more than $90,000).

Subcontract amounts are committed and become unavailable as soon as they are committed. The profit or loss on them is incurred, so there is no need to mix the amounts with the costs being recorded monthly to manage the work. The reason is simple: If the project costs reflect the $5,000 made on the electrical subcontract example, the job will look better than it is by that amount. Theoretically, the project could do $5,000 worse on other work before the system would indicate costs are not meeting the estimate for the entire project. If each line item stands on its own and if subcontract and material prices are considered spent as awarded from a cost control system perspective, the process is simplified and the tool is easier to use.

9.15 WORKING WITHOUT INFORMATION

If cost control information isn't accurate or isn't on time, don't use it. Bad information is worse than none. In these circumstances the alternative is much better: Trust your gut feelings. If contractors spend any amount of time in the field, they probably have as good a feel as necessary to make judgments on whether things are going well or not. And they should act on those judgments. If a contractor doesn't get to the field, they should talk to the people who do and find out what their gut feelings are. The people who have the best perspective on performance are those who visit the field periodically. Full-time field people usually have an optimistic view of how things are going and/or are hesitant

> *Your gut responses will be right most of the time, and you should rely on them until you can establish an effective cost control system.*

to report bad news. If a contractor doesn't get a good feeling about what they are hearing, they need to go to the field and find out for themselves. It's time to use all of that experience and know-how that's part and parcel of being a contractor. What is sensed on field visits is important, and contractors should follow their instincts even if the paperwork reveals the opposite. Is the site noisy enough? Is there enough excitement? How does it feel to you? Your gut responses will be right most of the time, and you should rely on them until you can establish an effective cost control system.

9.16   SUMMARY

It seems almost silly to say that all contractors who fail to profit do so because they didn't make money in the field. But it ceases to be funny when you realize that more than half of the distressed contractors that we worked with didn't know they were losing money until it was too late to do anything about it. Not evaluating contract profitability by project, by month, and by activity is one of the most serious and most avoidable causes of nonperformance. Cost control is fundamental to the business of managing construction for a profit, which is the contractor's and/or top management's primary function. There is one thing worse than losing money on construction, and that is losing money and not knowing it.

> *There is one thing worse than losing money on construction . . . losing money and not knowing it.*

CHAPTER REVIEW QUESTIONS

1. What are common excuses that contractors state as being the reason for not reviewing the accounting numbers each month?
   a. They do not have enough time.
   b. They do not have the proper financial training.
   c. They do not have the ability to manage the accounting function.
   d. All of the above
2. What commonly happens to the estimates of percentage of completion for a group of projects?
   a. They average themselves out, with some being high and some low.
   b. They are estimated to be too low.
   c. They are estimated to be too high.

3. Under GAAP rules, who is responsible for the accuracy of the estimate of the cost to complete the work?
   a. The independent accountant
   b. The contractor
   c. The auditor
   d. All of the above

4. A work in progress schedule requires which input columns?
   a. Contract price
   b. Amount billed to date
   c. Direct cost to date
   d. Estimated cost to complete
   e. All of the above

5. The primary problem in construction accounting is that an independent accountant has no way to verify the accuracy of what?
   a. The billing period in which liabilities are recorded
   b. The number of projects currently in progress for the construction company
   c. Original estimates of total project costs submitted as part of the construction company's bid package
   d. The figures for percentage of completion

CRITICAL THINKING AND DISCUSSION QUESTIONS

1. What is the significance of a construction business handling a lot of money that is not actually theirs?

2. What do you consider the primary purposes for being in business and why?

3. Discuss the percentage of completion method of accounting for partially completed work, and what do you consider the advantages and pitfalls?

4. Why is it important to a construction business to account for the profit or loss of all of their work in progress?

5. What do cost control systems actually control, and how important is the field input of data to such a system?

NOTES

1. Block, S. B., and G. A. Hirt (1989). *Foundations of Financial Management* (p. 26). Homewood, IL: Irwin.

2. Slavin, A., I. N. Reynolds, and J. T. Miller (1972). *Basic Accounting for Managerial and Financial Control* (2nd ed., p. 3). New York NY: Holt, Rinehart and Winston.

3. Halpin, D. W. (1985). *Financial and Cost Concepts for Construction Management* (pp. 17–18). New York, NY: John Wiley & Sons.

4. *Audit and Accounting Guide, Construction Contractors* (pp. 89–93), May 1, 1992. American Institute of Certified Accountants.

5. Ibid., pp. 86–89.

6. Ibid.

7. Op. Cit., Bersch (1973), Do you manage or just watch? *Constructor*, (September).pp. 1–11.

8. Op. Cit., *Audit and Accounting Guide* (1992), pp. 91–93.

9. Op. Cit., Bersch (1973), pp. 1–11.

10. Op. Cit., *Audit and Accounting Guide* (1992), p. 93.

11. Op. Cit., Ryan, K. N., Lecture May 7, 1993.

12. Remmen, A. (1977). *The Contract Bond Book.* Cincinnati, Ohio: National Underwriters, pp. 27–28.

13. Op. Cit., *Audit and Accounting Guide* (1992), pp. 90–93.

14. Op. Cit., Remmen, A. (1977), pp. 214–247.

10

EQUIPMENT COST MANAGEMENT

10.1 OWNERSHIP COSTS

The use and ownership of equipment affects every type of construction company, and accurately accounting for actual costs is a complex process and has caused problems for many. The process of evaluating and accounting for all costs associated with equipment usage and ownership is complicated by the fact that some costs do not appear on invoices and, in fact, aren't incurred in the course of operations. These costs, while not easily recognized, will eventually have to be paid. Although various formulas are used to cope with equipment costs, the total costs of ownership are sometimes misunderstood. All costs, including those hidden, must be recognized and planned for in advance.

Contractors face many equipment concerns such as whether to buy or lease, which equipment to invest in, and when to invest. Equipment-intensive contractors, like road builders, landscapers, sheet metal contractors, and others have to make these decisions often, and they have a long-term and profound effect on success. Other contractors have the same concerns, but less effort may be put into these decisions because equipment isn't a big part of their business.

Debt service and maintenance costs may exist for contractors who own very little equipment, but these costs usually exist in a big way for equipment-intensive contractors. This chapter discusses problems primarily facing equipment-intensive construction businesses, but the principles apply to all construction companies that own equipment.

10.2 HOW MUCH TO OWN

The first step in controlling equipment costs is to control the amount of equipment owned. The decision to purchase new equipment is made for basically two reasons: to replace aging equipment or for expansion. Both reasons are certainly valid business reasons. Nevertheless, both reasons must be considered judiciously because the company is usually committing a great deal of its money with limited assurance of future work.

In replacing equipment, management must weigh very heavily whether the new equipment is really more productive than what they have, and if so, by what margin? Is that difference worth the investment? If the

> *The first step in controlling equipment costs is to control the amount of equipment owned.*

existing piece of equipment is a maintenance headache, should the company invest in a complete reconditioning and get three or four more years out of it or replace it? Are the next two to five years of work a certainty? Is the marketplace growing or shrinking, or is it likely to change soon in either direction? Of course, no one can answer these questions with certainty.

The decision to purchase new equipment, which must produce profits over a period of years or be a financial liability, is not an easy or simple one. The decision to buy means taking on additional costs and creating a necessity to get at least enough work to keep the equipment busy. Too often, contractors "want" to buy newer and bigger equipment rather than "need" to buy it. When they need a replacement, some assume that bigger is better.

10.3 REASONS TO BUY

The decision to buy additional equipment for expansion is usually made for one of two reasons: New work is already contracted and there is no owned equipment available to do work; or the contractor is in an expanding marketplace and wants to have the equipment on hand to do the anticipated greater volume.

If the marketplace is getting stronger and is growing, it may be reasonable to assume that the company will get their share of the growth and, therefore, greater volume. The problem with buying or committing to more equipment in advance of getting the work is, as already mentioned; the company *must* then get more work just to keep the equipment busy and to justify the investment.

> *Equipment can run a contractor instead of the other way around.*

10.4 COMPETITIVE POSITION

One of the difficulties in getting this new amount of work is that the company's competitive position relative to their marketplace may not stay the same when the marketplace grows. Their regular competition may also have a bigger appetite and may be going after the work more aggressively than they did in the past. Another situation that often develops when a marketplace gets stronger is the influx of outside competition. When out-of-area contractors are drawn to a strong or growing marketplace, they need to get a foothold and often bid very tight to get the first job or two. Local contractors may react by bidding even more aggressively. New equipment becomes a real burden when it forces the organization to go after a greater amount of work at a time when they must bid work at a lower markup in order to get it.

Even if a company has contracted for more work than their current equipment will bear, the conditions of an expanding marketplace still make purchasing risky. Once the existing work starts to finish up and the equipment used on those jobs becomes available, the company may have idle machinery. Increasing inventory of equipment should be very carefully thought out, not only as it relates to the work, but also as it relates to the marketplace, the company's competitive position, and resources to do the additional work at a profit. Many successful contractors have loaded up on equipment in good times only to be forced into severe difficulties by the very same equipment when the marketplace went back to normal. The equipment begins to run these contractors instead of the other way around.

10.5   CALCULATING EQUIPMENT COSTS

The subject of calculating and accounting for owned equipment cost is one that is ignored by some organizations or lost sight of by others who believe their accountants are taking care of it. To estimate work and bid a job, a contractor needs to know exactly how much equipment is going to cost per unit time, and has to include all maintenance and replacement costs to do that. A contractor must have this information to accurately price the work at a profit and to know whether ongoing jobs are profitable. The basic concept for costing of equipment is quite simple, but calculating it can be another story.

The basic objective of operating a business is to produce net income, which results from receiving more from a customer for services rendered than the total expense of producing the service. As assets such as equipment are used in operations, they lose part of their service value, or "depreciate." This is an element of expense, called "depreciation expense."[1] A portion of the actual cost of the asset (equipment) expires in each accounting period during the useful life of the equipment. This periodic cost requires no periodic cash outlay, but, nevertheless, is a continuous expense of operating the business.[2] There are a number of methods that may be used in calculating depreciation by GAAP rules, but all are based on the purchase cost of the asset (equipment). Although depreciation is an estimate,[3] it cannot be based on replacement cost of the equipment, which presents a serious problem for the equipment-intensive contractor.

10.6   TIME AND USAGE

The cost of owning equipment is a function of both time and usage. Some equipment may be busy all the time under normal one-shift-per-day conditions, for example, a rock crusher. It can work for 12 months or seasonal businesses. The company owns the rock crusher when it starts on a particular job, and the machine is intended to operate all day, every day. It was purchased new. The direct costs to the company during the first month of operation, assuming mobilization is charged separately, are fuel, insurance, and regular maintenance. These costs are fairly easy to track monthly because they will be incurred during each month of operation. However, as there may be major maintenance as the project progresses and spare parts required, an allowance for this must be included in the equipment costs.

The allowance for maintenance and parts is an estimated cost that should be tracked and corrected occasionally to reflect what actually occurred. The allowance should be treated as an actual (if not incurred) cost because in several months, the crusher may need new belts and bearings, and the cost for these are not correctly chargeable to the month incurred. The belts and bearing were consumed over several months, and this maintenance cost should be charged in a timely manner by estimating it in advance in an attempt to reflect reality. These very real costs must be captured in a timely manner to account for the true cost of ownership of the equipment, but because some of these costs are yet to be incurred, they can not be arrived at from the company's accounting records. They must be estimated.

Estimated regular and extraordinary maintenance costs are very real and should be included in the unit cost of owned equipment because they are necessary to keep the equipment operable and in the condition it was when it was purchased. If these costs are not charged to the unit costs and charged against the work, then the cost of some maintenance will come out of profits. The estimated cost should be applied to cost control the same as incurred costs to ensure monthly costs are accurate. Estimated maintenance costs should be updated periodically (usually annually). If the equipment will need major overhauls such as engine replacements after two or three years, these costs have to be factored into the estimated maintenance and accounted for from the first month the equipment is put into service (Figure 10.1). If this is not done, the company will

© iStockphoto.com/publicimage

Figure 10.1   As Long as the Equipment Is Working, Most Equipment Accounting Methods Capture All the Costs Associated with Ownership. A Prudent Contractor Accounts for Costs Whether or Not the Equipment Is Active.

be overstating their real profit by not charging the wear of the equipment to the jobs that caused it. It creates a false economy.

10.7 REPLACEMENT COSTS

In order for equipment-intensive contractors to enable their companies to replace their machinery, they must charge a replacement cost to the unit costs. The replacement cost should not be confused with the purchase price used in depreciation calculations. Furthermore, these charges should be applied to the cost control system hourly, weekly, or monthly as a true (actual), if not incurred, cost. The replacement value is calculated by determining the useful life of the equipment and estimating the replacement cost, less salvage at the end of its useful life. The replacement cost is divided by the useful life to get the monthly cost that will be incurred. In this calculation the number of hours, weeks, or months the equipment is anticipated to work per year is used. The replacement cost represents the cost to the company of using up or consuming the piece of equipment. If an organization uses the purchase price and not replacement costs as many do, and divides by the useful life, it will not collect the replacement cost of the equipment during its use because of inflation.

A contractor may ask, "Do you really want me to charge my clients for next year's inflation when I'm only working for them this year?" The answer is, "No. You can always pay for it yourself." If an equipment-intensive contractor does not charge replacement cost, they are paying for the privilege of being in business. They are consuming equipment at rates that are intended to repay the company for the amount they paid for the machinery. However, when the equipment is replaced, for example in five years, the company will need more than what was originally paid for the equipment to replace it. To be self-sustaining, an operating business must replenish or regenerate itself from operations and total cost of ownership. It is an economic reality that inflation is a cost of doing business and for an equipment-intensive contractor, this means charging replacement cost to the unit costs of equipment as it is used.

Economic analyses supporting a decision on equipment replacement are aimed at determining the equipment replacement interval that will yield the maximum on the equipment investment. The period of equipment ownership that yields the maximum profit on the equipment investment may be considerably shorter than the economic life of the equipment. Equipment ownership costs, as the term implies, represent the cost of owning the equipment. Although these costs are usually prorated on an hourly basis for estimating and accounting purposes, they represent costs that would be incurred whether the equipment is actually used or not.[4]

10.8 EQUIPMENT COSTS CHARGED TO PROJECTS

The purpose of charging all equipment costs to the jobs and applying these charges monthly as costs are incurred in a cost control system is to give the contractor a realistic picture of whether the company is making or losing money in time to do something about losing situations. Equipment is not an overhead cost any more than moving equipment back to the yard stops ongoing costs.

Consider the heavy equipment contractor during a slow period. The company can either charge more of the equipment costs to fewer jobs or consume the difference. During a slow period, there will be no fuel cost, and maintenance can be suspended, but the insurance cost goes on as does replacement cost. Replacement is a function of usage, age, and obsolescence. It is often obsolescence that causes replacement, so the timing of replacement isn't only affected by usage. If a contractor believes that downtime will extend the useful life of the equipment, then they can adjust the replacement cost as long as they factor in an amount for deterioration from storage and non-usage. Deterioration can be a costly factor because most construction equipment wears better in use than out of use.

> *Taking work just to break even is rarely justified, except for survival.*

10.9   IDLE EQUIPMENT

The alternatives open to a contractor whose equipment is idle because of an inability to capture profitable work are not encouraging. To take work in other geographic locations (see Chapter 4) or on a tight schedule isn't good business because the company takes on too much risk just to keep the equipment working. To take highly competitive work just to break even is rarely justified, except for survival. Liquidating some equipment is an alternative, but must be considered in the context of the overall business, including new work anticipated. There is seldom a profit to be made in liquidating used construction equipment, although liquidation can reduce losses caused by the ongoing cost of idle equipment. Leasing out idle equipment is a favorable alternative, but this is usually difficult to do if there is a general slowdown in the market.

If nothing can be done to mitigate the loss from idle equipment, you might consider leaving it on the last job it worked on so that it does not go unnoticed. This serves as a reminder that the equipment is idle, and management, who should have anticipated when the equipment would be free, will be encouraged to get everyone talking about where it should go next. It also tends to discourage project people from always asking for more equipment than they really need.

Because most idle equipment is not charged to projects, it is possible for all jobs to be showing a profit when the real picture for the company is not as good as the paperwork is showing.

10.10   CASH FLOW

Cash flow is equal to the sum of earnings (after taxes) and depreciation.[5] Taxes on earnings are paid in cash, reducing the company's cash flow, but depreciation is a non-cash expense and thus contributes to cash flow.

The majority of the cost of equipment ownership does not occur concurrently with the equipment's usage; thus, equipment-intensive contractors usually have a positive cash flow that can be mistaken for profit. Even if an organization accepts these concepts and accounts for all of the costs as previously described, the company's cash flow will be greater than their real profit. If the replacement charges and extraordinary maintenance charges are not accounted for and actually placed in reserves, then funds won't be there when needed. During slow periods with a lot of idle machinery, an equipment-intensive contractor could be showing losses on all jobs, but still have a positive cash flow and is therefore able to weather the storm well. If the company uses some or all of the funds reserved for equipment replacement and if they are not replenished out of future profits, there isn't going to be enough money to replace the equipment when the time comes.

> *... positive cash flow ... can be mistaken for profit.*

Because most equipment is purchased not with cash but on credit, the equipment is expected to be paid off from future work. The example used earlier was for equipment purchased for cash because it makes the concepts easier to relate to and follow. For equipment purchased on credit, there is a slight change in the proposition. Interest costs are added to the formula as an expense similar to fuel and insurance. Since both interest and principal must be paid concurrently with usage, cash flow during slow periods is affected. During idle time or when losses occur for any reason, there may be a loss with a negative cash flow. Depending on the length of a slow or losing period, there may not be enough cash flow to make the equipment payments.

10.11 EQUIPMENT OBSOLESCENCE

Equipment-intensive contractors have another exposure in the equipment area that is not as apparent and often not planned for, and that is equipment obsolescence. Companies incur a great deal of cost in replacing equipment as it wears out. Broken-down equipment delays jobs, hinders progress, and costs money.[6] Additionally, obsolescence can occur well before machinery reaches its useful life. The productivity of equipment dramatically affects the profitability of equipment-intensive contractors and is part of their competitive edge. Equipment productivity is critical to making a profit and to bidding and getting the work. As newer and more productive equipment comes into the market and a contractor's competitors buy it, the contractor can be forced into equipment replacement earlier than planned just to remain competitive. Equipment obsolescence prior to useful life is a difficult issue because it is almost impossible to predict and consequently to plan for. It is therefore a risk of doing business for equipment-intensive contractors.

> *The construction industry is mobile, and unanticipated competition can come from anywhere.*

10.12 EQUIPMENT OBSOLESCENCE CASE STUDY

This case study is about a well-established sheet metal contractor with aging duct-making machinery that faced new and unexpected competition from a start-up contractor who had the latest technology equipment. The productivity of the new equipment allowed the start-up contractor to bid lower on every job of any size that came out during their first year in business, until the established contractor had hardly any work. The established contractor decided the only way to remain in business was to replace their equipment with the more productive machinery their competition was using. Like most contractors, they had not reserved money for equipment replacement or, for that matter, even accounted for it. The company's current financial statements reflected a bad year because of the new competition, and the established contractor was unable to purchase the new equipment because they could not secure the financing. In fact, the new equipment was so expensive that their last five years' total profit wouldn't have paid for it.

The contractor was very aware of potential equipment obsolescence and the benefits of new technologies, keeping abreast of the latest developments in their field such as computer-operated duct-fabricating machinery. In fact, they knew that eventually some or all of their equipment would need to be replaced but felt that they still had a lot of good years left in them. However, replacement costs were not included anywhere in their cost accounting. By ignoring the real cost of replacing its equipment, the company was enjoying a "false profit." By accounting for realistic replacement reserves, the contractor would have seen that the real profits weren't what was thought. Additionally, had the contractor considered obsolescence, it may have been possible to plan for continual upgrading of equipment or at least measure how far the company was falling behind technology and quantify the risk and cost to the business.

Contractors must understand that competition is a strong force in the construction industry, including specific construction disciplines and subdisciplines. The notion that the established contractor could not have foreseen the eventuality of the start-up company entering the marketplace is not the issue. New businesses with better ideas are a reality in any industry, existing competition may gear up and tool up at any time to increase their market share, and out-of-town contractors are always on the look out for new areas to expand into, particularly if existing competition appears weak or less productive. An equipment-intensive construction enterprise that does not concern itself with the effect of obsolescence on their business, and with remaining at least as productive as industry averages nationwide, is a business that is at great risk.

Simply reserving the anticipated cost of "keeping up" is not enough because a company may not be able to gear up fast enough if the competitive balance shifts rapidly. It is necessary to also spend the reserves and keep up with national standards, not just local standards, because the industry is mobile and unanticipated competition can originate from anywhere.

Figure 10.2 Early Example of Heavy Construction Equipment.

10.13 REPLACEMENT COST INCURRED DAILY

The entire future replacement cost of equipment, including the costs due to inflation, obsolescence, and wear, necessary to remain in business will become due whether or not it is accounted for or reserved by contractors (Figures 10.2 and 10.3). Replacement cost is a very real cost of doing business, and is a cost that is incurred each day, not only at replacement time. Following is a simplistic example.

A contractor decides to go into the dirt-moving business and buys a $100,000 bulldozer: How will they account for the ownership of this piece of equipment in years to come? Let's say they buy it for cash from personal savings and that it will last for five years, at which time it will be worn out. For this example, we'll assume zero salvage value. There are a number of ways to account for depreciation, and we'll select straight-line depreciation over five years or depreciation of $20,000 a year.

If the new business recovers all other usage costs during that five years and charges only $20,000 depreciation in their accounting for equipment ownership, where will they be in five years? They will no longer have their $100,000 because they spent it to buy the bulldozer in the first place. They won't have the bulldozer because it is worn out and has no salvage value (use 8, 10, or 12 years if you like; it comes out the same), and they won't have a job because they don't have the piece of equipment. What happened to the $100,000? It has been consumed by the business. Sure, they made profits during those

© iStockphoto.com/john330

Figure 10.3 Powerful Tools And the Forerunners of the Machinery That Helps Build the Country.

years and the depreciation allowed them to have $20,000 of the profit without corporate tax, but the $100,000 was after-tax dollars. They spent the company's profits on salary and operating costs. A new bulldozer today costs more than the $100,000 it cost five years ago, say, $150,000. To stay in business they need to borrow the $150,000. In this situation, the contractor is not only short the $100,000 that they started with five years ago, but they are also in debt another $150,000.

While the example is simplistic, it provides good food for thought. Of course, replacement costs to be incurred five-plus years in the future would be calculated at the present value of the future cost, and tax considerations would impact the calculations. Nevertheless, replacement costs will become due at some point and will almost always be more than the original purchase price, so one cannot rely on allowable depreciation alone to accurately account for the cost of ownership of equipment. If the company buys equipment on credit and replaces it with credit, as most contractors do, the company will go deeper into debt by at least the rate of inflation the longer it remains in business.

> *Equipment replacement costs won't be incurred for five or more years, which are not a big problem unless you intend to be in business longer than that.*

10.14 SUMMARY

When a company does not account for the real replacement cost of equipment, profits are exaggerated, which gives a false picture of where the organization is and certainly of where they are headed. While the IRS does not allow funds reserved for equipment replacement costs to be tax free, they are clearly a cost of doing business. There is a lot of debate on this subject from accountants and tax experts, and it would be beneficial to unite and determine what the optimal method is for all parties considered.

Industry economic conditions raise the question of whether a fleet of equipment is a liability or asset. Ignoring the real replacement cost of equipment necessary to remain in business can influence an organization to operate in a false economy, go further into debt over time, and for many contractors create serious long-term financial problems. Costs that are incurred and due in the current accounting period are no more real than costs that will definitely be incurred in the future and will become due in a subsequent accounting period; it is just harder to recognize them and account for them. In the example provided, equipment replacement costs won't be incurred for five or more years, which is not a big problem unless you intend to be in business longer than that.

CHAPTER REVIEW QUESTIONS

1. What are the main reason(s) for deciding to purchase additional equipment?
   a. New work is already contracted, and there is no owned equipment available to do work.
   b. Because of an expanding marketplace, the contractor desires to have the equipment on hand to do the anticipated greater volume.
   c. None of the above
   d. a and b

2. Which of the following is a component of the total cost of equipment?
   a. Replacement cost
   b. Maintenance cost
   c. Purchase price
   d. All of the above

3. What are some of the risks for equipment-intensive contractors?
   a. Equipment obsolescence
   b. Equipment idling
   c. Mistaking positive cash flow for profit
   d. All of the above

4. Replacement costs are incurred when?
   a. Yearly
   b. Quarterly
   c. Monthly
   d. Daily

5. Which of the following methods will most closely capture total costs of ownership?
   a. Calculating the entire future replacement cost of equipment, including the costs due to inflation, obsolescence, and wear
   b. Depreciation
   c. Expert estimation
   d. None since it is impossible to capture total costs of ownership

CRITICAL THINKING QUESTIONS

1. What do you consider the criteria for which equipment to own for any particular business?
2. Does the equipment it owns affect a construction company's competitive position and if so, how, and if not, why not?
3. Where does a construction company charge idle equipment to and why does that matter? Discuss the issue.
4. Does the age and/or productivity of a piece of equipment impact a construction company's competitive position and why?
5. How would you go about determining the useful life of a piece of equipment and how would you estimate its future replacement cost after its useful life?

NOTES

1. Slavin, A., I. N. Reynolds, and J. T. Miller (1972). *Basic Accounting for Managerial and Financial Control* (2nd ed., p. 3). New York, NY: Holt, Rinehart and Winston.
2. Ibid., p. 3.
3. Ibid., p. 14.
4. Nunnaly, S. W. (1977). *Managing Construction Equipment: Solutions to Problems* (pp. 213-214). Upper Saddle River, NJ: Prentice-Hall, Inc.
5. Smith, K. V. (1979). *Guide to Working Capital Management* (p. 5). New York, NY: McGraw-Hill.
6. Diamant, L., and H. V. Debo (1988). *Construction Superintendent's Job Guide.* (p. 19). New York, NY: John Wiley & Sons.

11

OTHER INDUSTRY CONCERNS

11.1 INTRODUCTION

This chapter covers a number of miscellaneous areas of concern for construction organizations. Each of them has caused performance deterioration or worse for a number of companies and any of them un-addressed can induce financial distress. If an organization is suffering from any of the other elements of contractor failure addressed earlier, these miscellaneous areas of concern exacerbate and amplify the problem.

The categories are presented in no particular order, as the degree to which an organization experiences the concern dramatically influences the impact on the company. Many of the concerns may appear minor if an organization is not suffering from them. However, any one of them can impair a construction company, and if they get out of hand, they can do great harm. Some of these are very difficult to discern in an organization and easily exist for many years unrecognized by management. For that reason this chapter bears rereading periodically by managers responsible for the success and well-being of a construction organization.

11.2 GROWTH AND RISK

In the volume-driven industry of construction that thrives on growth, there are difficulties even among the older and well-established firms during growth or downsizing. The words "growth" and "growing" recur in the study of the management of risk in the construction business because the business risks in construction are simply greater during growth phases and equally so during downsizing.[1] A construction company must be managed well to be successful, and in the best of times there is risk. A rapidly expanding construction company magnifies its risks even if it is closely and intensely managed. There is nothing wrong with building a bigger and bigger business. That is the American dream. But the increase in risk in the construction industry

> *. . . increase in risk in the construction industry from growth alone cannot be understated . . .*

from growth alone cannot be understated and should not be overlooked. Downsizing voluntarily or when market driven similarly magnifies risk.

11.3 MARKET DRIVEN

The ideal construction company would be organized to be market driven and not volume driven. It would strive for carefully planned growth but be prepared to level off or fall back on volume if the marketplace tightens or shrinks. It would use its markup flexibly as a competitive tool but never take break-even work just to maintain volume. In a tightening market (greater competition for the same work) or in a shrinking market (less work available), the ideal construction company would bid more competitively than it would in a better market but concentrate on making at least a minimal profit on less work. It would have some "flexible overhead" built into the organization that could be cut immediately and would not hesitate to cut permanent overhead when downsizing is necessary.

The ideal construction company is willing to get small again to survive. Each down cycle will pass and they will be ready for the upswing, but only if they come through intact (Figure 11.1). The large failure rate in the construction industry is driven in part by construction enterprises that push full speed ahead during weak or down markets with desperate bidding in an attempt to capture work that their competition needs as badly as they do.

11.4 CONTROLLING THE NEED FOR VOLUME

Overhead costs are difficult enough for a contractor to control when the company is not growing, but in a growing organization they pose two dangers. Because an organization cannot add a half-person or a half-piece of equipment, they are forced to put on overhead costs during growth in larger amounts than perhaps they would like. This can cause losses until the company grows into the overhead. Herein lies the double problem: First, companies will lose money, or

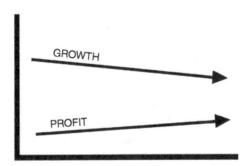

Figure 11.1   When Sales Drop in a down Market, It Is Still Possible to Downsize And Manage Profit.

profits will be reduced for a length of time because of sudden increase in overhead to accommodate growth; and second, needing additional volume creates stress and strain to resources in order to cover the increased overhead.

As an organization attempts to increase market share, price will suffer because it is always necessary to make at least temporary price concessions to take the market share away from competitors unless there is an exceptional boom market, which is usually temporary. Boom markets attract the attention of out-of-area companies who, when they move in, also make price concessions to take the work away from local contractors and get a foothold. While construction companies may not make a conscious decision to lower their price when they must have the added volume or new work, it is exactly what occurs. And when price suffers, it is usually for all the new work, not just part of it, so the company ends up needing even more volume than originally planned because margins are suffering. This can easily lead to a downhill profit spiral during rapid growth and often does because as an organization gets stretched, there is little time for anyone to see the problem coming.

> *As an organization attempts to increase market share, price will suffer.*

The additional growth then requires more overhead, creating temporary losses and the immediate need for even more volume. This spiral can result in serious financial stress.

Rapid growth will also put a strain on the company's key people and systems, and sustained growth doesn't allow for a reasonable training period. Of even greater concern, continued growth doesn't give a contractor a chance to test new people or systems before the next new people are put on and new systems are added. If performance or profit starts to deteriorate during growth, it is almost always discovered after additional volume and people are taken on, and corrective measures are more difficult with everyone already stretched out. Overworked managers will be coping with the largest volume the company has ever handled. Unprepared contractors will pursue continuous growth with no performance measurements established.

11.5 RATE OF GROWTH

If a company's market is not growing, growth is obviously more difficult, but in a reasonable market construction companies are almost always growing at some rate. In our experience growth for a construction enterprise at a rate of more than 15 percent in a year should be considered substantial. Sustained growth for more than a couple years compounds quickly. At 15 percent a company doubles in five years and triples in seven, at 25 percent it doubles in three years and triples in five, and at 50 percent it doubles in 20 months and is five times larger in four years.

Growth requires more resources in the way of people, systems, and money. Success is measured in an organization's ability to find the necessary qualified people, have appropriate systems in place in advance of expansion, and finance the growth. Rate of growth obviously impacts the likelihood that an organization will be able to bring qualified resources to bear on the new work in a timely fashion. The alternative is to expect existing resources to do more, but few construction organizations are known for having underutilized resources or "bench strength."

As volume increases, the expanded company is untested as an organizational unit. The only reasonable test is for the new organization to operate profitably and smoothly for a minimum of a year. Sustained growth creates a situation in which if the test proves unsatisfactory, new growth has already been added during the test year and the company is looking at a second bad year before they can roll back to their proven size and proven team. For many it is too late to retreat and recover.

Incremental growth instead of sustained growth may seem unnecessary, even unnatural, but it is the best way to control the inherent risk in growth beyond 15 percent. Constant testing of small growth allows a company to reevaluate and recover if they encounter an unproductive time period. It is prudent risk control.

In sustained growth a company grows beyond its people and systems so often that they never really have the same organization long enough to truly test it, and are at constant risk with an ever-changing team. In some cases it's just a matter of time.

> *Incremental growth . . . may seem . . . unnatural, but it is the best way to control the inherent risk in growth beyond 15 percent.*

11.6 FLEXIBLE OVERHEAD

"Flexible overhead" is a new concept for the construction industry. The marketplace is so unpredictable and affected by so many variables that it is difficult to accurately forecast for even a few years. If enterprises cannot be sure of a stable or growing market while causing or allowing the businesses to grow, they can control risk by putting on overhead (to deal with the growth) that can easily be removed if the market turns sour. With some of their overhead flexible, the company does not become a slave to their volume and can fall back if necessary and concentrate on profit.

The method is to use temporary employee services for some clerical administrative and accounting functions. Use short-term rentals for some office and field equipment and short-term office leases, even temporary trailers, during growth stages until a new plateau of volume can be reasonably assured. Even management people can be brought on with specific company growth and performance goals associated with their continued employment. This creates challenges for new people and refocuses the real risks associated with growth for existing management. The practices have been successful with established companies and start-up firms in construction and other industries.

There are costs associated with flexible overhead as short-term leases and rentals cost more than purchased equipment, temporary employees may cost more, and efficiency may suffer slightly. But the reduction and control of risk is well worth the modest additional expense. An added benefit is that existing managers get involved and excited about this prudent, realistic, and businesslike approach to growth. They can easily discern the positive impact flexible overhead has on their job security.

Flexible overhead may create cramped quarters and less creature comforts (privacy, plush offices, and the latest technologies), but those who use it to control risk during incremental growth phases say they sleep a lot better when they get home at night. Putting on permanent overhead in a fickle market is just too dangerous. Most who have tried the flexible overhead approach have been impressed with it to the degree that they put permanent overhead on even slower than would normally be considered safe. They are committed to keep some portion of their overhead flexible at all times as a hedge against a market slump, and that portion seems to grow as they realize how easy and economical it is. The modest added cost is not unlike an insurance premium for protection from a known and measurable exposure. Companies that embrace flexible overhead manage their profit and not their volume.

Ideally, the flexible overhead concept can be managed to prepare the construction enterprise to do 25 percent less volume in any given year just as it is also prepared to do 25 percent more and have no increase in risk either way. An organization skilled in flexible overhead has learned how to gear resources up and down temporarily and more quickly and economically than an average construction company can secure permanent resources. It is realized that this is a departure from the accepted norm, but it is clearly the profile of the successful contractor of the future.

> . . . *flexible overhead . . . is a departure from the accepted norm, but it's . . . the profile of the successful contractor of the future.*

11.7   MOBILITY OF THE INDUSTRY

There were always peaks and valleys in the construction marketplace like there are today; however, historically when things were bad in contractors' normal work areas, they had to stick it out and do the best they could. Twenty or thirty years ago contractors stayed in their own backyards. Construction companies generally worked a lot closer to home because their equipment wasn't as mobile as it is today, short-term leasing wasn't as prevalent, and travel and relocation were more difficult. But when things were good, construction companies and all their competitors had a seller's market. Because contractors weren't that mobile, they didn't go into new areas in great numbers and impact the market, so there was greater opportunity for substantial profits during good times.

The expression, "they took the good with the bad," is appropriate here. The good years allowed for great earnings, and in a more conservative era some of this "excess" would be put away as reserves against the lean years. Reserved or not, when a seller's market developed, contractors were able to generate substantially greater percentages of profit than they can today under any market conditions. The major reason is increased competition, which is a result of greater mobility.

Today whenever a good market develops anywhere in the country, out-of-area contractors compete for a portion of it, invariably preventing a seller's market from developing and in many cases actually driving down prices. There are very mobile nationwide contractors that follow the good markets, and they are supplemented by contractors from any area of the country where there isn't enough work and who are willing to travel.

11.8 DIMINISHED PROFITS

The net effect of greater construction industry mobility is that peaks are taken out of the various markets while the valleys remain. The opportunity for really big years is substantially reduced, and the

> *Limited profit margins require contractors to grow with prudence, test as they go, and be prepared to withdraw from bad decisions.*

average profit in the industry has diminished over time and shows every sign of staying down. Ease of mobility nationwide and internationally will continue to maintain competitive pressures, which in turn keeps prices down.

What this means to the average contractors is that without the prospect of boom years to make up for bad ones, they must take care to control their valleys. With typically limited cash reserves, contractors can ill afford to increase risks without controls and must manage their businesses cautiously, if not defensively. Limited profit margins require contractors to grow with prudence, test as they go, and be prepared to withdraw from bad decisions.

11.9 EMPLOYEE BENEFITS AND COMPENSATION

After the concepts of flexible overhead and peaks and valleys, the subject of employee benefits and management perks ties in well. The general and administrative costs of doing business are as necessary to the running of a construction company as are the costs of concrete and steel. Controlling these costs is imperative. Overhead costs deriving from benefits and perks must be treated judiciously, and the best way to do that, particularly in good years, is to be conservative in preparation for bad years.

The discussion of bonuses is an intricate part of the management of overhead costs. Performance bonuses are quite common in the construction industry; however, many firms mismanage them. To be effective, performance bonuses must be part of a carefully considered compensation plan, which is known and understood by all of the participants. Random, unorganized, and separate deal bonuses, common in the construction industry, cause more problems than what they are worth. Some companies have even fallen into the trap of giving bonuses each year regardless of company performance. These bonuses tend to become regarded by employees as part of their wages unless the bonuses are tied to the performance of the employee or the company; preferably, they should be both.

Random or unorganized bonuses add overhead costs spontaneously and haphazardly, and the benefits from them diminish very rapidly. To be effective, bonuses must be part of a formal, overall compensation plan. They must be tied to each individual employee's performance, the profitability of the entire job, and the success of the entire company.

The cost of bonuses or unrealistic compensation packages put in during good years has accelerated the stress on many companies when lean years hit. Luxury automobiles, club memberships, and pleasure trips are near and dear to anyone's heart and are common perks for hard-working managers in many construction enterprises. The biggest problem with these overhead expenses is that the costs to maintain them keeps going up and up while loyalty and motivation resulting from them goes down as they become expected or are taken for granted by employees and managers.

A company car is a tremendously valuable perk, and often it is given in lieu of a $2,000 or $3,000 raise in a particular year even though it's worth much more than that. The problem is that two or three years later it is almost always taken for granted by the valued employee who now only understands one thing: that he or she is underpaid by $2,000 or $3,000 compared with somebody else. Giving such perks is hard to avoid because so many organizations are doing it, but there is great value in having the highest paid people around with no perks. A construction organization with high pay and few perks is usually able to keep their employees longer and has little trouble getting new ones to quit their lower paying jobs to come to work for the company, because in most cases they have lost sight of the real value of their perks. It's cheaper in the long run and more professional.

Another problem with perks is they are often selected by the contractor or management as something they themselves value and would like to have or already have. Most perks are offered to employees without an alternative for different benefits or wages and are therefore never refused, and common courtesy demands gratitude so the giver has no real test of the level of appreciation or the value the employee places on the perk. For example, a conservative employee provided a company car that is replaced every two to four years may very well prefer to drive an older more economical vehicle and have the cost of the company car added to his or her salary.

Our experience suggests that employee perks provided by well-meaning construction companies are universally undervalued by employees who, when they learn of the cost to the company, almost unanimously state they would prefer to revive the value in compensation. Unfortunately, it is difficult for a contractor to test this among their own personnel because employees do not wish to appear ungrateful, and experience suggests that employees respond very differently to their employer than they do to an anonymous questioner.

11.10   MOTIVATION AND LOYALTY

The efficiently managed construction company is businesslike and professional and attempts to make its long-range plans around key positions, not key people.

Many closely held construction companies, particularly smaller and mid-size organizations, operate in a family or club-like atmosphere, and many contractors believe it generates loyalty and longevity. There is no credible research to suggest differently, but the practice is fairly expensive, and in the current working environment in this country it is becoming more difficult to maintain. With the mobility and lifestyle of middle management, two-career families, and a shrinking work force, people are changing jobs with greater frequency than ever before. Job security and company loyalty aren't the top concerns of today's work force. They are being replaced with quality-of-life issues and job satisfaction. Changing jobs or relocating without a job to go to is more common than remaining 15 or 20 years with the same company, which employees tended to do in years past.

Managers should look back over their company's history and recall who the key players were five or seven years ago. For many this is an ever-changing scene, and it will change even more in the future. A close-knit group working in a club-like atmosphere may appear efficient, but if the players are ever changing, then a portion of the money spent on creating the family atmosphere might better be reserved for recruitment and paid out in training replacements for key people. Increasingly the efficiently managed construction company today is businesslike and professional, has a certain amount of internal competition among managers, and attempts to make its long-range plans around key positions, not key people.

11.11   INTERNAL COMPANY DISPUTES

The construction industry may be one of the largest industries in the country, but it is made up for the most part of hundreds of thousands of small- and medium-size companies, most of which are closely held or family businesses. Internal disputes among the management of closely held construction companies are not uncommon and have created discomfort and disruption for a number of contractors and their families. In a high-risk, low-margin industry where businesses are often operated at high levels of intensity and energy, some conflict can be expected.

For whatever reason, personal problems within small- or medium-size, closely held construction companies seem to cause more problems than they should and have been seen to affect company performance and profits significantly. Some managers expect more from family members and are more tolerant of non-relatives. Further, the succession of leadership in closely held companies is often not as smooth as it should be, even in some cases when a great deal of planning has gone into it. Conflict is an area of concern for owners of closely held businesses, and it is imperative they be aware of it and provide for open and honest communication of all parties involved at all levels within and outside the organization. If continuous disharmony is affecting performance after management has attempted to resolve it, professional intervention is essential. Unresolved friction can fester and degenerate beyond repair and has caused untold personal anguish and distress. It can distort a successful organization and cause serious underperformance.

11.12   DEBT

A construction company uses credit in many ways. Most construction organizations need some or all of the following: a line of credit with a surety company to secure payment and performance bonds on projects; secured loans to purchase the equipment needed to do the work; and bank credit to fund working capital either on a seasonal or as-needed basis or to fund growth. Bank credit is part of the everyday needs of a construction contractor, and the management of that credit requires skill and attention. Arranging credit is not an event; it's a process. The process is another of the many things a contractor needs to manage to remain successful.

A contractor should become concerned with any unplanned or unscheduled use of credit. It is too common for construction organizations to borrow working capital unexpectedly and not fully understand why the need arose. Management should be aggressive in determining why the money is needed today when there was no anticipation of the potential need last month. If an organization is borrowing working capital unexpectedly, it signals the need for better cash flow planning or that profitability is falling off. A construction business operating without good cash flow planning is out of control for the simple reason they never know when they are going to run out of money.

A large line of credit is no substitute for cash flow planning. A company that doesn't borrow at all must still have cash flow planning. But for a company that borrows some or all of its working capital, cash flow planning is critical. Not only is there the possibility of running out of both cash and credit, there is the added interest cost to be considered. Without cash flow planning an organization will spend more on interest than they will with cash flow planning, even if the planning isn't done well.

Cash flow planning must be included in all decision-making processes. The first consideration in business is profit. The second consideration is cash flow, and some managers believe the order should be reversed. Every business decision must include the answers to these questions: Will this change, decision, or project create a cash outlay? Will it provide an influx of cash? How soon will this occur and at what risk?

Borrowing should never be delegated to middle managers. It is important to the security of the business and is intricately involved in the process of controlling the risks the company faces at any given point in time. Borrowing should be controlled by top management through careful planning that takes into account the amounts and timing of credit use and addresses the sources and timing of payback. The need for any unplanned borrowing should be cause for great concern because either the cash flow plan is wrong or it isn't working. In either case, new planning is required and the new planning on short notice should be undertaken with the same diligence as the original effort.

11.13 BUSINESS PLANNING

Strategic planning, both short-range and long-range, is not addressed in any formal way by many contractors. That's not to say contractors don't express their objectives and the plans they have to achieve those objectives. What it means is that they don't have a written guide. They know their objectives but have no detailed plans to achieve them. When they are without at least informal plans with some details, businesspeople must react to whatever comes their way. It becomes very difficult to retain control, set the direction, and measure the progress in that direction.

Short- and long-range formal written plans are the tracks on which a company runs. They make managing a construction business so much easier. It is surprising that contractors don't devote more time to developing and following strategic short- and long-range plans because doing so is such an effective tool. The time spent in the planning comes back to the organization with incredible interest and dividends of time saved.

> *Short- and long-range formal written plans are the tracks on which a company runs.*

Planning should be done at a time set aside for just that purpose and outside the mainstream of daily activity.[2] The owners and key managers of the firm should discuss and evaluate their individual and corporate goals and see how they fit. Not everyone wants to go to the same place, perhaps not even in the same direction. All the company's resources should be realistically evaluated and measured against short- and long-range goals to see if they fit. By establishing clear goals and directions that are understood by everyone concerned, meeting these objectives becomes easier if only because everyone is thinking along the same lines and looking in the same direction.

Things certainly don't always go as planned, but much of a contractor's business future is actually within his or her control. When things change, plans can be reevaluated and altered. This way, the organization is not simply reacting, but acting in an organized fashion. The plan provides a measure of movement. The importance of short- and long-range plans that are formally written and detailed cannot be overstated. A construction company needs at a minimum a one-year detailed, or hard, business plan and three years of soft, or flexible. plans. A two-year plan may be sufficient for a smaller or newer company. Long-term business planning is the ultimate risk control tool available to a contractor. Companies new to the planning process find that they go about the job of managing with less frenzy and more confidence and purpose.

11.14 RECOMMENDATIONS

It is recommended that construction professionals be careful, prudent, business-like, and professional as they manage their construction enterprises. They should treat their employees, associates, and other parties in the construction process as they would like to be treated. They should develop a business that they can truly manage with confidence so that some of the enjoyment comes back into building.

CHAPTER REVIEW QUESTIONS

1. How does increased volume negatively impact companies (circle all that apply)?
   a. Sudden increase in overhead to accommodate growth
   b. Decrease in employee salaries
   c. Stress and strain on existing resources
   d. No negative impact because of the increased profits
2. Growth requires more what?
   a. Money
   b. System
   c. People
   d. All of the above
3. Flexible overhead would consist of which items(s) (circle all that apply)?
   a. Yearly building lease
   b. Temporary staff
   c. Purchased equipment
   d. Short-term leased equipment
4. To be effective, bonuses should include or be part of what?
   a. Company vehicle
   b. A formal compensation plan
   c. Vacation trips
   d. All of the above
5. A construction company at minimum should have a _____-year detailed, or hard, business plan and a _____-year soft, or flexible, plan.
   a. one, two
   b. one, three
   c. three, six
   d. five, ten

CRITICAL THINKING AND DISCUSSION QUESTIONS

1. What are risks and advantages to rapidly growing a construction enterprise?
2. What do you consider constraints to such growth and how would you overcome them?

3. How you would create some flexibility in the overhead structure of a construction organization.

4. How would you develop policies that create and maintain motivation and loyalty among your employees and what would be the costs associated with these activities?

5. What is the significance, if any, to business planning for a construction company and how much time do you thing should be devoted to such activity?

NOTES

1. Lewis, V. L., and N. Churchill (1983). "The Five Stages of Small Business Growth." *Harvard Business Review* 61(3):30–50.

2. Gibson, G. E., Y. Wang, and M. P. Pappas (2006). "What is Preproject Planning, Anyways?" *Journal of Management in Engineering* 22(1):35–41.

PART 2

The second section contains the "elements of success": appropriate practices that assist in avoiding or counteracting the elements of failure presented in earlier chapters and advanced strategies and concepts about how to manage risk and profit in the highly risky and volatile construction industry. Construction accounting basics are addressed briefly as an introduction to the sections on financial management, which present high-level strategies and methods concerning the financial aspects of construction businesses of all sizes. Chapters 12–16 address complex accounting and financial issues in a straightforward manner for the non-accountant construction professional and demonstrate how to use and develop data to manage the business, not just report about it. Changes in project delivery methods are discussed along with the impacts they are having on the industry. Advanced strategies about how to capture work in a changing market are outlined and detailed.

12

FINANCIAL MANAGEMENT ISSUES

The construction industry is challenging in many aspects, especially in the area of financial management. The challenges facing construction companies that make them susceptible to billing and collection challenges, cash flow problems, unexpected profit declines, credit squeezes, and worse are discussed. Following are the key issues that form the core of the financial management challenge in the construction industry:

> The construction industry is challenging in many aspects, especially in the area of financial management.

- The managers of construction companies are normally operations people who are in touch with the pulse of the business at the *project* level, but often lack the knowledge, training, and information to truly understand the business at the *company* level.
- Financial and accounting processes in the construction industry are unique and complex, and are rarely in any organized format.
- Due to the of the lack of meaningful, industry-specific training, the construction company's financial people often don't have the ability to get beyond mere number-crunching to make significant contributions to management.
- External certified public accountants (CPAs) are often not specifically trained in the specialized knowledge that is unique to the construction industry, and therefore cannot be truly helpful to management.
- This industry is heavily dependent on outside credit (primarily banking and bonding), but few managers are adept at dealing with those creditors.
- To obtain accurate financial information, operations people must be closely involved with the financial process (especially work in process calculations), and they are often unaware of the importance of their involvement, and lack the training and awareness to do it well.

12.1 KEYS TO SUCCESS

Construction companies can be successful, or unsuccessful, at any size and at any stage of their existence. While the previous list describes some of the challenges related to the business of construction, the following addresses the keys to success:

- Outside factors (like competition or the market for construction services) are very important, but the decisions made by management are more critical to success than any external occurrences.
- No company can "win" consistently by playing defense (reacting); the company must play offense.
- Operations people can be effective at getting construction projects built and at working with numbers, accounting issues, and paperwork.
- Financial folks can be effective in providing the right information to the right people in a timely manner and have enough insight into the construction operations to provide meaningful strategic input.

The science of running most types of companies—manufacturing, distribution, retail—is fairly well understood (though there are obviously competing concepts). Even the art of running companies in those industries is fairly well explored. The science and art of running a construction company is not well explored, well documented, or perhaps even well understood. Much of the mystery surrounding the management of construction companies involves the financial aspects of the business (Figure 12.1).

12.2 WHAT FINANCIAL STATEMENT ARE SUPPOSED TO CONVEY

Contractors can choose between various potential "profit models," which is true for all businesses. The profit model is a targeted combination of volume, gross profit, and overhead. For example, if a company does $3 million in volume (revenue) at 20 percent gross profit and carries an overhead of $40,000 per month, it will make $120,000 in profit for the year. This is one potential profit model for this company. Another profit model would be if this company would also make $120,000 in profit by taking on a volume of $7.2 million with a 10 percent gross profit and a $50,000 monthly overhead, or doing $43.2 million in volume with a 4 percent gross profit and $134,000 per month in overhead.

There are an infinite number of potential profit models, and probably just as many reasons why to choose one model over another. The size and type of projects the company normally undertakes, availability of good work in the market, the gross profit margins anticipated to earn on that work, and basic overhead structure are all major considerations in this choice. Other factors that come into play in choosing one of these models are how strong the company is financially

Figure 12.1 Time Devoted to Financial Management Is Critical to a Contractor's Success.

and how much risk they are willing (and able) to take. In the previous example, it would obviously take more cash, and involve more risk, to do the $43.2 million in work with $134,000 per month overhead than it would to accomplish the other two models.

The exact profit model being implemented may not be clear to everyone, but most managers have some idea of the ranges of results they anticipate: "Next year we'll probably get between $x million and $x million in work, our typical margins are between x percent and x percent and overhead runs about $x thousand per month."

The primary task in reviewing financial statements is to determine whether the company is on track to accomplish the expected profit model, and whether the company has enough financial resources to continue on the projected course. Financial statements need to be analyzed at various levels, such as individual balances, ratios, and other key indicators.[1]

12.3 THREE MAJOR FUNCTIONS

The three factors described in the profit model correlate with the three major functions in a construction company: get the work (revenue), do the work (gross profit percent) and manage the business (overhead). For each of these major functions, the company is somewhere on a continuum with "capacity" on one end and "stress" on the other. By spending money, capacity can typically be added to any of these functions. By growing (increasing usage) without adding capacity, the movement will be toward the stress side of the spectrum.

The capacity end of the spectrum is more comfortable and has brighter potential (there is room to expand), but most managers believe it's less profitable. After all, if there is capacity, there may be payment for idle (or underutilized) resources.

The other end of the spectrum—stress—is not as comfortable, more problems arise, and management can't maneuver as well into the future (management feels confined because they are already doing as much as they can do). However, the stress side is usually considered to be more profitable than the capacity end of the spectrum because it is perceived that higher output can be achieved from existing resources (no resource is underutilized). Having significant amounts of underutilized resources will erode profits; however, remaining at the stress end of the spectrum erodes profits when profitable new work can't be undertaken (due to full capacity) and costly mistakes abound.

Consider the three primary functions in a particular company. As far as the "get the work" function is concerned, is the company closer to the capacity end, or is it more toward the stress end? If management decides right now to go after some new market, do they have excess capacity in their business development or estimating teams?

Do they have capacity in terms of "do the work," or is the company stressed? The symptoms of companies that are stressed in this function are essentially various forms of job breakdowns. Interpretation of key financial statements can provide further clarity and understanding of why this may be occurring.

12.4   FINANCIAL STATEMENT BASICS

Consider what the primary financial statements, the "balance sheet" and "income statement," are supposed to indicate. Most construction people are far more comfortable with the income statement, also referred to as a profit and loss (P & L) statement. Therefore, in this section we put extra emphasis on becoming more comfortable with the balance sheet.

The objective of the financial information process is to obtain quality, timely, and accurate information regarding two extremely fundamental issues about the company:

1. How healthy the company is financially (the balance sheet).
2. In what direction the company is heading financially and how fast it's moving in that direction (the income statement).

A company that's financially healthy has adequate financial resources to do the amount and type of work they are doing with their existing overhead structure. The financially healthy company has enough cash to meet their obligations (they are liquid), including enough slack that a hiccup or two isn't a serious problem. Put another way, the company has adequate financial resources to accomplish management's intended profit model.

> *A company that's financially healthy has adequate financial resources to do the amount and type of work they are doing with their existing overhead structure.*

A very financially healthy company would have more financial resources than they really need to do the good work that's reasonably available in the market. It may be unwise to have excessive financial resources in a company because more assets are at risk than is necessary, but for now we will look at this condition as a good thing. Thus, the bills can be paid and more work can be pursued if desired—and still have resources left over.

Quite simply, a financially *un*healthy company doesn't have the financial resources to do what they are currently doing or have been doing. A single hiccup can push the company over the edge. Cash flow difficulties are the classic symptom of this condition. In this condition, financial resources are stressed or stretched. It is absolutely critical to know where the company stands health-wise.

> *…a financially unhealthy company doesn't have the financial resources to do what they are currently doing or have been doing.*

The second thing necessary to understand is where the company is headed, and that's where the income statement comes in.

The income statement provides several critical pieces of information: whether the company is growing in volume either quickly or slowly, shrinking in size either quickly or slowly, or staying about the same. Think of this change in volume as "speed." The income statement also indicates whether the company is making money, losing money, or breaking even. Think of profitability as "direction."

If these statements can be easily understood, the balance sheet and income statement work together to provide the critical information needed to manage successfully.

12.5 BALANCE SHEET

The balance sheet (Table 12.1) is a snapshot of a company's financial status on one particular day—normally at the end of a month, quarter or year.[2] The basic equation for the balance sheet is:

$$Assets - Liabilities = Equity$$

Table 12.1 A Sample Balance Sheet Format

ABC CONSTRUCTION BALANCE SHEET December 31, 20xa		
ASSETS		
CURRENT ASSETS		
Cash		$ 166,685
Contract receivables		1,347,367
Inventory		169,463
Prepaid expenses and other current assets		172,240
Costs and estimated earnings in excess of billings on uncompleted contracts		455,605
	TOTAL CURRENT ASSETS	2,311,360
PROPERTY AND EQUIPMENT		444,492
OTHER ASSETS		22,225
		$ 2,778,077
LIABILITIES AND SHAREHOLDERS' EQUITY		
CURRENT LIABILITIES		
Current portion of long-term debt		$ 44,449
Accounts payable		711,188
Accrued liabilities		147,400
Billings in excess of costs and estimated earnings on uncompleted contracts		341,703
	TOTAL CURRENT LIABILITIES	1,244,740
LONG-TERM DEBT, less current portion		238.915
		1,483,655
SHAREHOLDERS' EQUITY		
Common stock, no par value, 1,000,000 shares authorized, 100,000 shares issued and outstanding		100,000
Retained earnings		1,194,422
		1,294,422
		$ 2,778,077

Without getting into some of the more esoteric accounting definitions, "assets" can be defined as what a company has or possesses, "liabilities" can be defined as what a company owes, and "equity" (or net worth) is the difference between the two.

The best way to understand a balance sheet may be to consider a personal financial situation for a moment. If one is putting together a personal balance sheet as of some specific date, the first step would be to add up all assets: cash in the bank, investments, vehicles, home, etc. The next step would be to list all liabilities: home mortgage, vehicle loans, personal loans, credit card balances, etc. The amount by which assets exceed liabilities is net worth.

The amount of net worth is a measure of personal financial health, and so is the makeup or "quality" of that number. Many of these assessments may be intuitive; for example, cash and publicly traded investments are more liquid than other assets. Other assets, like a home, may have more stable values and are more readily saleable than art collections or furniture. Having long-term debt in the form of a home mortgage or a vehicle loan is less risky and less expensive than maintaining outstanding, current credit card balances.

If someone assembled their personal balance sheet, they would have a net worth "gauge" and would also be able to judge the overall quality of that net worth amount. Then, at the end of the next year of making some money, spending some money, making purchases, selling stuff, and saving a little, they can put together a new personal balance sheet.

They can now compare their updated net worth amount with their previous net worth balance, and can compare the quality (liquidity, stability of value) of their new balance sheet with the quality of their prior balance sheet. They would now know two important things: where they currently stand financially and how their actions of the prior year affected their financial health.

By taking a more personal perspective, the balance sheet and its uses can be understood better and extended to an organization's use of this information.

12.6 THE HOLDING TANK CONCEPT

Besides using the balance sheet to monitor a company's financial health, this statement can also be used in the accounting process as a "holding tank." To demonstrate this concept, we need to reconsider the income statement for a moment.

In order to build an accurate income statement, revenue and expenses must match in the same period. If the revenue from a project was placed only in the current period, and the related expenses were placed in the following period, the income statement will be misstated for both periods.

Although the logistics of matching are sometimes difficult, the accounting concepts are straightforward: If transactions are incurred that belong to a future event, the proper action is to defer (push out) those transactions until that event occurs. If the event has already occurred, but all of the transactions haven't yet come in, the proper action is to accrue (pull in) those future transactions into the current period.

An easy example to illustrate this concept is a small job done on a time and materials (T & M) basis. Let's say that a company does some work on the last day of the month and goes back the next day to complete the work, which happens to be the first day of the following month. The job then gets billed on that day, but costs—parts and labor—were incurred on the last day of the previous month. If the accounting process being utilized does not "hold" these transactions in the balance sheet, the records will show costs without any revenue in the first month, and revenue with incomplete costs in the second month. The company's results, income for both of these months, are distorted.

The accounting system needs to hold the costs from the last day of that first month in the balance sheet as work in process, so that the first month's income statement shows absolutely no effect from this work. The costs from the first month will be deferred into the second month, where they will meet up with the billing along with the rest of the costs to form a complete financial picture in a single period in the second month.

Another approach if most of the work occurred during the first month and only insignificant costs were incurred on the first day of the following month is to leave the costs in the first month and accrue the billing and the additional costs from the following month, in which case the transaction will fall entirely in the first month.

When events overlap the end of any period, it is necessary to identify all of the revenue and costs of that transaction. Then, it is necessary to choose which period they belong to, and use the balance sheet as a holding tank to account for the costs and revenue in a single period.

12.7 ASSETS

As previously described, assets can be defined as what a company has or possesses. On a company's balance sheet, assets are typically broken down into three sections:

1. Current
2. Property and equipment (or fixed assets)
3. Other

12.8 CURRENT ASSETS

Current assets are either cash or resources that will be turned into cash within a year. On a balance sheet, the current assets are normally presented in the order of decreasing liquidity, so having larger balances "higher" in the list is better. These are the most common current assets:

1. Cash (or equivalents like money market accounts or CDs)
2. Contract (or accounts) receivable (which should include retentions)
3. Other receivables (refunds and employee advances are common examples)
4. Inventory (materials bought into stock to be used in later projects)
5. Prepaid expenses (something paid for in the current period, but that won't be used until the next period—like rent or insurance)
6. Underbillings (revenue in excess of billings)

It is helpful to consider these balance sheet items in terms of their quality. For example, cash and receivables are obviously the best current assets to have. Other receivables, though usually small, are typically okay (meaning they are collectible). Receivables from the company owner(s) are frowned upon by outside creditors and should be avoided as much as possible.

Inventory is an acceptable current asset as long as the balance is an amount that will be turned quickly in the normal course of business and if the company routinely buys materials into inventory and then issues those materials to projects. Inventory is generally not okay if the balance consists of unused materials, especially unusual items that were dragged back into the yard from the last job, hoping they will be used in the future.

Prepaid expenses are usually acceptable assets, and relatively insignificant but should be minimized, especially since the bonding company will be using the statement. We have found it bad practice in most cases to pay for expenses in advance.

In considering the quality of balance sheet items, think about inventory and prepaid expenses in light of the holding tank concept. Both of these current assets consist of costs that are being pushed into future periods. The costs have already been paid, but they're still hanging around, waiting to be dropped into the income statement as they get used up.

Underbillings are often the most dangerous and controversial of the current assets. Staying with the holding tank concept, underbillings are future job billings that have been pulled into the current period. Underbillings are a calculated amount, not an actual bill. If a job is going to be billed in the next period, and the accountants are just accruing (pulling in) this billing for matching purposes, the underbilling is probably fine. If this job was not billed because there's a problem on the project, accruing the underbilling may result in a current asset that has no value, and the income will be overstated.

> *Underbillings are often the most dangerous and controversial of the current assets.*

12.9   PROPERTY AND EQUIPMENT

The property and equipment section of the balance sheet contains the company's "fixed assets." Fixed assets are tangible assets that have an extended life, usually three years or more, like vehicles, buildings, equipment, and furniture.

These assets are depreciated each year and are stated on the balance sheet at their depreciated original cost, not their "fair market value" (fair market value is what the asset is worth if a willing seller sells it to a willing buyer). Here are a few issues to consider regarding fixed assets:

- Items that are capitalized for book purposes are also capitalized for tax purposes.
- Depreciable life is set by the IRS for tax purposes, but the company needs to determine appropriate life for book purposes (they may be longer or shorter than tax lives).
- Most companies have different depreciation schedules for book and tax.

- A capitalization policy should be established and then left alone until the business changes significantly. For example, a company with $1 million in revenue usually has a lower capitalization threshold than a $10 million company.
- Items that can't be listed and described individually because they will be difficult to identify later when they are removed should not be capitalized (note that "desk" and "truck" are not adequate descriptions).
- The depreciation schedule should be reviewed annually to make sure that any items that have been disposed of have been removed even though they are fully depreciated.

12.10   OTHER ASSETS

The last section in the asset portion of the balance sheet is "other assets." These are simply assets that are not current and are not a part of property and equipment.

Some of common examples of other assets are:

- Long-term deposits, like a lease or utilities, but usually not plan deposits, which are short-term
- Cash surrender value of life insurance
- Assets held for investment: land, stock investments, an ownership interest in another company
- Intangible assets like goodwill or covenants not to compete (which typically result from the acquisition of a company)
- Long-term deferred costs like loan acquisition costs or organizational costs

The important issues related to other assets are to make sure that they're classified correctly. If the asset should be considered to be current, make sure it's classified that way, since that's advantageous, and that the other assets are easily identifiable to outside creditors. The key is to maximize working capital and be wary if financial statements show all of the other assets on a single line unless they're very small. Creditors may want to see them separately, and that may be to the company's advantage.

12.11   LIABILITIES

Liabilities are typically split into two sections on the balance sheet—current and long-term. The definition of current liabilities mimics that of current assets: Amounts that will be paid in the next year.

Here are some common examples of current liabilities:

- Accounts payable
- Accrued liabilities (or expenses)
- Current portion of long-term debt
- Bank line-of-credit

- Income taxes payable (if a C corporation)
- Deferred income taxes (if a C corporation)
- Overbillings (billings in excess of revenue)

Accounts payable are the bills received from vendors and subcontractors for job costs and overhead expenses. The common accounting challenges regarding payables are to make sure that all of the payables get recorded in the right period, including subcontractor billings, and that retentions payable are recorded as payables as well.

The concept of accruing expenses (liabilities) was discussed in the holding tank section. For all intents and purposes, accrued liabilities act like payables except that there is usually no invoice. When accountants close an accounting period, they fine-tune the matching of expenses in the period by doing things like calculating accrued payroll. For example, if there were two work days in the last week of the previous month that were paid in the first week of the following month, the amounts paid (i.e., payroll taxes, sales taxes, bonuses, profit sharing contributions, and other expenses) are accrued from the future into the current accounting period. If the company has expenses that are paid on a periodic basis, often annually or semiannually, the accountants can accrue an estimated amount on a monthly basis so that each month is depicted more accurately, and the company does not experience a wild fluctuation during the month in which the big annual expense hits.

The current portion of long-term debt represents the principal portion of term notes that will be paid in the next year. Typically, this amount is determined by using amortization schedules that break out the principal and interest portions of loan payments. The principal balance that will be paid with payments beyond the next twelve months is classified as long-term. The liability that appears on the balance sheet is the principal only. The interest expense will be recorded as the payments are made.

If the company has a line of credit with a bank, the amount available to borrow on the line is not recorded. When the line is used, the amount borrowed is recorded as a current liability.

Long-term liabilities usually consist of long-term debt or capital leases. These reflect the principal portion of the company's debts that will be paid in periods beyond the next year.

12.12 EQUITY

Equity is typically the shortest section of the balance sheet. The total of all the balances in the shareholders' equity section is also referred to as the company's "book value" or "net worth." Following are some of the common examples (these are for a corporation):

- Common stock
- Additional paid-in-capital

- Treasury stock
- Retained earnings

In many companies, common stock represents the original amount that the shareholders invested to capitalize the corporation at its inception. If the corporation sold additional stock after its inception, that amount is included in this total as well. If the shareholders sold stock outside of the corporation (person to person), there is no change on the corporation's books. Often, the common stock balance is a nominal amount that has no particular significance and seldom (if ever) changes.

Treasury stock results from the purchase by the corporation of the stock belonging to former shareholders. As mentioned earlier, if shareholders sell their stock to other people outside of the business, the equity section of the corporation is unaffected. When the corporation purchases the stock, the purchase amount is recorded as a deduction from the company's equity, and this (negative) treasury stock balance remains in the equity section until the treasury stock is retired.

If a corporation purchases stock from a shareholder using a note payable (buy back the stock now, then pay out the cash over time), which is often the case, the balance sheet gets hit really hard. Equity is decreased by the amount of the purchase, and liabilities are increased by the same amount (if there wasn't a down payment). If the company relies on credit, like banking or bonding, this can be a dangerous and very detrimental transaction because of the impact on the debt to worth ratio (a key ratio tracked by most creditors).

Retained earnings are, by far, the most active balance in the equity section of the balance sheet. Retained earnings are the cumulative sum of all income earned by the corporation since its inception, the net of any losses or dividends. Negative retained earnings are referred to as an "accumulated deficit."

Distributions are routinely paid to shareholders of S corporations to pay income taxes at the very least, and S corporation distributions may also play a significant role in income tax planning.

12.13 INCOME STATEMENT

The income (profit and loss) statement is the more user-friendly statement for most people with the typical basic equation[3] shown in Table 12.2.

Revenue − Cost = Gross profit
Gross profit − General and administrative (G&A) Expenses = Income from operations
Income from operations − Other expenses = Net income

In the construction industry it is common to refer to sales as *"revenue."* Revenue is the primary accounting challenge in the income statement, because, in construction, revenue is an amount that is calculated based on the progress on each job. Revenue is not the same as billing, and it's certainly not the same as cash collected.

Table 12.2 A Sample Income Statement

ABC CONSTRUCTION STATEMENT OF INCOME AND RETAINED EARNINGS Year Ended December 31,1996	
CONSTRUCTION REVENUE	$ 8,056,423
COST OF CONSTRUCTION	6,654,605
GROSS PROFIT	1,401,818
GENERAL AND ADMINISTRATIVE EXPENSES	1,087,618
INCOME FROM OPERATIONS	314,200
OTHER EXPENSE	(24,169)
INCOME BEFORE INCOME TAXES	290,031
INCOME TAX EXPENSE	(116,012)
NET INCOME	174,019
BEGINNING RETAINED EARNINGS	1,020,403
ENDING RETAINED EARNINGS	$ 1,194,422

Costs and expenses are synonymous terms, but costs are typically at the project level, while expenses are usually at the company level. Both of them can also be seen as assets that can no longer be used (cash that is no longer available). Other income and expense items are typically those that are not really a product of the construction process like interest income and expense, investment income, rental income, and gains and losses from sales of assets.

12.14 FINANCIAL STATEMENT SETS

On an annual basis, most contractors have a set of financial statements prepared for them by an outside CPA firm. These sets usually vary in format from the internally produced financial statements in several ways. Most notably, they have additional information (especially notes) and less account detail.

A complete financial statement set contains several elements (statements and notes), which are required to be presented according to generally accepted accounting principles (GAAP). The required elements and typical supplemental schedules of the financial statement sets for construction companies are noted in the numbered list soon to follow.

Although these supplementary schedules are not required by GAAP, creditors will normally require including them with statements. These schedules often contain information, particularly the job schedules, which may be considered sensitive if providing financial statements to clients or other parties besides bankers and bonding agents. Therefore, it's common to have a CPA prepare financial statement sets containing all of the supplementary schedules for the banker and surety and sets that do not contain the supplemental information for other parties.

Here is the typical breakdown of a full set of CPA-prepared financial statements:

1. Balance sheet (required by GAAP)
2. Income statement (required)
3. Statement of changes in equity, or retained earnings (required)
4. Statement of cash flows (required)
5. Notes to the financial statements (required)
6. Supplemental detailed schedules of contracts in process and completed contracts (along with an earnings summary that reconciles the job schedules to the income statement), detailed schedules of the cost of revenue and/or general and administrative expense sections of the income statement (not required by GAAP)

The statement of retained earnings is essentially a brief form of the statement of changes in equity. If the only equity transactions for the period consist of changes to retained earnings (profits or losses and dividends), but there were no other changes like sales or purchases of company stock, there will usually be a brief form and a statement of retained earnings at the bottom of the income statement. If there were transactions in the company's stock or other equity accounts, there will be a full version of the statement of changes in shareholders' equity in the financial statement set.

The statement of cash flows is interesting and useful but hard to interpret for most people. To add to the confusion, there are two versions of this statement: direct and indirect. The two versions are very similar in form. Effectively, the direct version has some additional details. The purpose of the statement of cash flows in either form is to reconcile net income or loss to the net change in the company's cash balance from one year to the next. Most people use this statement to obtain a few details that are not shown elsewhere in the statements: primarily the in and out activity in the company's property and equipment and debt sections.

Disclosure notes are required in statements and will typically contain:

1. Information about the company's activities; a general description of the business, where they work, who they work for, etc.
2. Information about the company's accounting policies
3. Additional information about balances that appear in the balance sheet like a breakdown of fixed assets by type, a breakout of retentions receivable and payable, terms of the company's debt, security, interest rates, payment amounts, and so on.

4. Information about common issues not otherwise displayed in the financial statements such as concentrations of revenue from certain customers or certain types of customers, amounts in backlog, terms of operating lease commitments and transactions with related parties.

5. Information that is not particularly common but could be vital to outside creditors like problems with lawsuits, guarantees the company has made (perhaps on the debt of the shareholders or another entity), claims on projects, or other serious issues.

Supplemental schedules are often a part of financial statement sets because key creditors rely on them for critical information. Most often, construction companies will include schedules of contracts in process and completed contracts, along with a summary schedule that reconciles those two schedules to the income statement.

It is also common to see the cost of revenue and general and administrative expenses broken down in detail as supplemental schedules, allowing the income statement to be very concise. There is a lot of variation in the presentation of these schedules (as there is in the presentation of job schedules), but they are typically far more "aggregated" (or summarized) than the detailed listings of costs and expenses as presented in internal statements.

12.15 SUMMARY

The two primary financial statements, the balance sheet and the income statement, are used to assess a company's financial health and the speed and direction of the company's movement.

The balance sheet is a snapshot of the company on a particular date, allowing for the evaluation of key indicators, like working capital and equity, along with evaluating the relative quality of the assets. In the accounting process, the balance sheet is also used as a holding tank to facilitate the matching of revenue and costs in the same period.

The income statement, which covers a period of time, is used to gauge the company's direction and speed. Growth and profitability are the key indicators. Every company's profit model—how much volume will the company perform at a certain gross profit while carrying a particular overhead—is played out in the income statement.

Different profit models involve varying levels of financial risk, and it's critical to understand the financial health of the company as portrayed in the balance sheet before selecting a particular profit model.

CHAPTER REVIEW QUESTIONS

1. Your company's year-end report is approaching. What can you do to look as good as you can for the annual snapshot?

    a. Minimize "junky" accounts
    b. Don't be overdrawn in your cash accounts
    c. Pay down lines of credit
    d. Delay major fixed asset purchases until after the year-end report
    e. Enthusiastically collect accounts receivable
    f. All of the above

2. When my CPA firm shows me a draft copy of our company's balance sheet, what am I really looking for?
    a. A large reserve of cash
    b. Accounts receivable to know how much I am owed
    c. Ensure that assets and liabilities are categorized correctly
    d. Costs and expenses that are categorized correctly

3. When my CPA firm shows me a draft copy of our company's balance sheet, what else should I be looking for?
    a. Assets, revenues, and net profit
    b. Assets, revenues, and expenses
    c. Revenues, expenses, and stockholders' equity
    d. Revenues, expenses, and net profit

4. The objective of the financial information process is to obtain quality, timely, and accurate information regarding which two extremely fundamental issues about the company?
    a. How healthy the company is financially
    b. If the company has any money
    c. In what direction the company is heading financially and how fast it's moving in that direction
    d. What is the company's fiscal year

5. What are common current assets?
    a. Other receivables
    b. Inventory
    c. Underbillings
    d. All of the above

CRITICAL THINKING AND DISCUSSION QUESTIONS

1. What are the core financial management challenges in the construction industry?

2. Discuss what financial statements are supposed to convey and why?

3. Explain the concept of capacity versus stress and why that is the case?

4. Explain the difference between the balance sheet and the income statement and what each is supposed to convey?

5. Explain the terms "equity" and "net worth" and what they mean to a company owner.

NOTES

1. Weygandt, J. J., P. D. Kimmel, and D. E. Kieso (2009). *Financial Accounting* (p. 22). New York, NY: Wiley.com.
2. Ibid., (p. 24).
3. Ibid., (p. 22).

13

FINANCIAL ANALYSIS AND INDICATORS

Financial analysis is the art and science of putting financial information into a format that will generate action. The science aspect of analysis involves calculating or accumulating a variety of indicators, which can be used to evaluate various financial performance areas. The art of analysis comes into play when various combinations of these indicators raise questions and suggest scenarios. The ultimate goal of financial analysis is to blend the observed facts together and convert these facts into key ideas, which can then be converted into actions by the managers of the company.

These ideas become condensed into phrases such as "we're growing too fast and that's putting pressure on our financial condition"; "our overhead is too high for our expected work next year"; "we don't have enough working capital to execute our plan"; or perhaps even "things couldn't be better, keep it up." It's far easier for managers to take actions with ideas like those in mind than to base their actions on various ratios.

Ideas and actions begin by gathering facts. Financial analysis is best accomplished by using three to five years of financial statements going through the following steps:

1. Calculate and track working capital.
2. Calculate a series of ratios and indicators that cover the four major analysis categories—profitability, liquidity, leverage, and financial capacity.
3. Note the trend of each category of ratios (for example, "liquidity is falling sharply" or "profitability is increasing steadily").
4. Compare the key ratios to any acceptable industry or trade standards that can be found. Good standards can be hard to find for various reasons.
5. Wrap the trends and industry comparisons together into phrases like "liquidity is falling sharply and is now well below industry standards" or "profitability is increasing steadily and exceeds the industry norms."
6. Consider all of the elements together in order to arrive at conclusions about the state of the company's financial health, the direction the company is

moving, any danger signs, and, most importantly, which of these signs are symptoms and which are causes.

> *Ideas and actions begin by gathering facts.*

13.1 WORKING CAPITAL

The financial analysis process begins with working capital. Managers of construction companies must ask themselves many vital questions, but none are more critical, and more overlooked, than the question of whether the company has enough financial resources to do the volume that the managers want to do (Figure 13.1). Working capital helps answer this key question.

> Working Capital = Current Assets – Current Liabilities

Working capital is found in the balance sheet and is calculated by subtracting total current liabilities from total current assets.[1] Creditors often adjust working capital in their credit calculations, usually disallowing or dropping what they consider unreliable asset categories or calculations. The basic equation for working capital is: Working Capital = Current Assets – Current Liabilities.

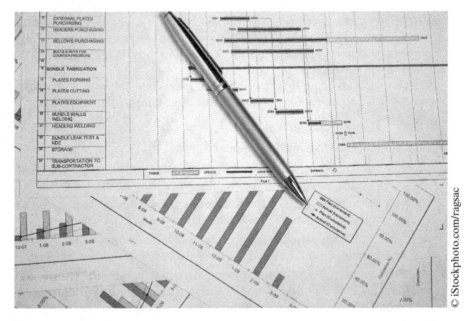

Figure 13.1 Financial Capacity And Strength Can Impact on Time Schedule Performance.

13.2 CALCULATING TARGET BACKLOG

Working capital can be used to determine how much backlog the company can carry without overly stressing their financial resources. This is done by applying a factor or multiple to the calculated working capital. A starting place for this multiple for most small- to medium-size trade contractors is 10. Multiplying this multiple times, working capital gives you target backlog. Larger companies or general contractors who subcontract a lot of their work will often use a higher multiple than 10. If a company tends to bill customers very late, for example, only after a project is completed or after substantiating the billings with invoice copies, it probably have to lower its multiple. Multiples of 7 or 8 are usually the low end of the range and 15 is on the high end, so 10 is a pretty good guideline for many companies.

A general guideline is that no more than half a company's backlog should be in one job. Another tip that may be considered is not to bid a job that has estimated labor costs greater than the company's working capital because it's possible for labor to go dramatically over budget, which could lead to financial distress. While it seems natural and desirable to take on larger work and to grow a company over time, prudence is appropriate. A company would benefit from the establishment of guidelines or policies about the amount and size of work that will be attempted while keeping in mind how much is too much.

> *A general guideline is that no more than half a company's backlog should be in one job.*

13.3 CALCULATING TARGET ANNUAL INCOME

Target backlog can be used to determine a target annual volume. Target annual volume is target backlog multiplied by work turnover. If the average job lasts two months, the company will turn work over six times per year. If the average job is one year in duration, the company will turn work over once a year. For example, if work is turned over three times per year with a planned backlog of 10 times working capital, the target annual volume will be 10 multiplied by three turns or 30 times working capital.

For most contractors, job length varies considerably. In this case, the typical or average duration would be used as a rule of thumb. Another method is to determine typical backlog and typical annual volume, then divide volume by backlog. There is no need to be overly concerned about the precision of these calculations and assumptions because they determine only a guideline.

In the next section is an example of how to calculate the target annual volume.

13.4 MAXIMIZING WORKING CAPITAL

Since working capital is total current assets less total current liabilities, the way to maximize working capital is by maximizing current assets and minimizing

EXAMPLE: WORKING CAPITAL

Facts:

Total current assets = $1,000,000
Total current liabilities = $600,000
Average job duration is three months (therefore annual turnover is times 4).

This is a small contractor who doesn't subcontract a lot of work, and bills on normal monthly cycles with normal retentions.

Solution:

Working capital is $400,000.
Using a times 10 factor target, maximum backlog is $4,000,000.
Maximum single job would be $2,000,000 (half of the $4,000,000).

Target annual volume would be $16,000,000: the target backlog of $4,000,000 times the estimated annual turnover of times 4.

current liabilities. The following are common transactions and their effect on working capital:

Increasing working capital:
1. Earning profits on jobs
2. Selling long-term assets (like property and equipment)
3. Borrowing money on a long-term note

Decreasing working capital:
1. Losing money on jobs
2. Buying long-term assets (like property and equipment) for cash
3. Paying off long-term debt

No effect on working capital:
1. Collecting receivables or paying payables
2. Borrowing money on a short-term basis, like a line of credit
3. Buying long-term assets on a long-term note (though there is some effect from down payments and the current portion of the note)

Most contractors like to buy equipment for cash if possible and want to pay off debt as early as they can. In concept, both of those things are good, but they'll both decrease working capital. Contractors also tend to avoid the opposite—selling equipment and adding new long-term debt; however, both of these are likely to increase working capital. The primary challenge is maximizing a company's working capital, which sometimes may seem counterintuitive.

An example that illustrates the importance of working capital follows. A third generation owner of a family construction business had taken over the company. Originally, the founder had purchased the property on which the company was located about 50 years earlier. During the second generation, the company had financial difficulties and remortgaged the business property in order to keep going. When the third generation took over, the company's goal was to pay off the property as quickly as possible, with the thought being that the business would have better future cash flow in the absence of the mortgage payment.

After a very profitable year the mortgage on the property was paid in full. However, after paying off the mortgage, the company had insufficient working capital, causing a severe drop in bond credit. Work slowed to a crawl as the company struggled to keep going and pay the bills. They ended up remortgaging the property for a larger balance than had been paid off so that the additional amount covered the losses in the interim, but the company was already fatally damaged.

There are numerous other examples, but the bottom line is that if a company spends too much of their working capital on buying equipment or paying off debt, they won't have the capital to do work.

13.5 LIQUIDITY

Liquidity measures are used to determine a company's ability to meet its current (and expected) obligations. These indicators are typically calculated by comparing assets that are already cash or about to become cash with the liabilities that need to be paid in the same time frame.

Some of the common liquidity measures are:

> Liquidity measures are used to determine a company's ability to meet its current (and expected) obligations.

1. Current ratio (current assets divided by current liabilities)
2. Quick ratio (cash + receivables divided by current liabilities)
3. Receivables to payables ratio (receivables divided by payables)

13.6 CURRENT RATIO

The current ratio assumes that all of the current assets will turn to cash in the coming year, and that this total is available to pay all of the current liabilities.

This is perhaps the most common of the liquidity measures used by nearly all creditors. It's often used in bank covenants, and many analysts look for a current ratio greater than 1.3 as a sign of good liquidity. Even small movements in this indicator are meaningful.

13.7   QUICK RATIO

Using all current assets in a liquidity calculation isn't very conservative, thus some use the quick ratio, which doesn't consider prepaids, inventory, or under-billings in the calculation. Essentially, the quick ratio determines whether cash plus receivables will adequately cover current liabilities.

Some prefer to use the quick ratio in conjunction with the current ratio, allowing the current ratio to measure the amount of working capital, and the quick ratio to measure the quality of current assets (since it disregards the less secure current assets such as inventory, prepaid expenses, and underbillings). A general standard is that if current ratio should be above 1.3, then quick ratio should be above 1.1.

By example, consider that a company's current ratio stays at 1.3 from one period to the next, but the quick ratio falls from 1.1 to 0.9. The amount of working capital may still be adequate, but more of that working capital is tied up in lower quality current assets. It is more challenging to pay bills the following month with inventory and underbillings than it is with cash and receivables. It is prudent to watch liquidity carefully, particularly if these two ratios vary greatly.

13.8   RECEIVABLES TO PAYABLES RATIO

The receivables to payables ratio is fairly simple: Divide the receivables balance by the payables balance. As an indicator, it varies dramatically from one company to the next, but it tends to be pretty consistent within a given company.

In order to understand movements in this ratio, it is beneficial to consider the different types of payables. Payables primarily contain invoices for materials, subcontractors, and miscellaneous job costs like equipment rentals. The general and administrative (G & A) component of payables (phone bills, rent, office supplies, etc.) tends to be fairly constant in most companies. Other costs, primarily labor, are not in payables because they're paid weekly in most companies or as they're incurred.

A company's receivables to payables ratio should be calculated for a number of periods to understand where it typically falls. Changes in this ratio can then be used to detect problems. If the ratio declines, it can signal falling margins or billing problems. If it increases, it may indicate collectability issues with receivables. It is possible that changes in this ratio are caused by changing conditions in the work or a problem that needs to be addressed.

13.9 LEVERAGE

Another key area for analysis is the concept of "leverage." The more debt the company has compared to its equity, the more leveraged it is.

An example of two companies is used to illustrate. Company One has total assets of $1,000,000, total liabilities are $500,000, and equity is $500,000. This company's debt to equity (or debt to worth ratio), the key measure of leverage, is 1 to 1 ($500,000 debt divided by $500,000 equity).

Company Two has the identical $1,000,000 in total assets, but with $800,000 in liabilities and $200,000 in equity. This company would have a debt to worth ratio of 4 to 1 ($800,000 divided by $200,000). Obviously, Company Two is at much greater risk of having creditors make demands it can't meet than Company One.

A business is able to grow faster using debt dollars rather than equity dollars because debt dollars are much more easily available. The amounts borrowed from suppliers and subcontractors alone can quickly dwarf the amount invested in a company plus the amount retained in earnings, especially for a young company.

> *The more debt the company has compared to its equity, the more leveraged it is.*

The age of the company, its historical profitability, and its growth pattern are the primary influences on the debt to worth ratio. Young companies, less profitable companies, and fast growth companies tend to have higher debt to worth ratios and, of course, high risk. For a company that has been around long enough to establish itself, most credit extenders will want to see a debt to worth ratio less than 2 to 1. This is a commonly tracked ratio and is often found in bank covenants.

13.10 FINANCIAL CAPACITY

A key evaluation area of a company is its "financial capacity." These indicators, which also may be referred to as "operational" measures, are almost always critical in determining the cause of problems and, of course, in preventing them.

Two common financial capacity measures are:

1. Revenue to working capital ratio (revenue divided by working capital)
2. Revenue to equity (revenue divided by total shareholders' or members' equity)

These two ratios behave similarly, though the working capital version is more volatile and potentially more useful for most contractors. Revenue to working capital is one of the essential ratios in measuring financial capacity.

One of the most critical issues a company faces is deciding how much work to do. If the company grows too fast where the revenue is high compared to working capital, there will usually be deterioration in liquidity measures, and leverage will typically increase. When a company grows quickly, all of the components in the company are working harder and are at much greater risk of breaking down. Falling liquidity measures and rising leverage are indicators that risk of breakdown is increasing. The increasing revenue to working capital ratio indicates the cause, which is, of course, growth.

For example, if a company's annual revenue is $5,000,000 and the working capital at the end of the year (either an average or year-end amount) is $250,000, the revenue to working capital ratio is 20 to 1 ($5,000,000 divided by $250,000). If a second company did the same $5,000,000 in volume, but only had $100,000 in working capital, its revenue to working capital ratio would be 50 to 1. The increased risk of trying to do the same volume with much less working capital is obvious

In the discussion on working capital, we considered a backlog multiple of 10 as fairly typical for most trade contractors (and smaller general contractors) and a typical annual work turnover (if average jobs are four months long) of 3. Using these two figures, the target revenue to working capital ratio would be 30 to 1 (10 times 3). If the target ratio is 30 to 1, and a company is at 50 to 1 (like the second company mentioned earlier), it is very likely stressing its resources and needs to slow down or bring in additional working capital. These concepts suggest that slowing down, or not taking as much work as possible, is a viable scenario.

The aforementioned first company with a 20 to 1 ratio and a 30 to 1 target has additional capacity for new work or new ventures. The most effective place to be in a financial sense is below the revenue to working capital target, but not too far below. The company that did $5,000,000 in volume with $1,000,000 of working capital, a 5 to 1 ratio, has way too much capital they are not using, and should consider other productive ways of putting it to use. Besides the fact that they are not using the additional capital, it is at risk to creditors as long as it's in the company. On a positive note, with a 5 to 1 ratio creditors would be very comfortable.

> *One of the most critical issues a company faces is deciding how much work to do.*

13.11   ADDITIONAL INDICATORS

There are several additional indicators that are important, but don't exactly fit into the previously mentioned categories:

1. Break-even point and profit planning
2. R Score
3. Change percentages

13.12   BREAK-EVEN POINT

A company's break-even point is calculated by dividing projected annual G & A expenses by expected gross profit percentage. Obviously the objective is not to just break even, but it is a useful exercise in evaluating changes in overhead, entry into new markets, and a variety of what-if scenarios. See the calculation in the featured example:

EXAMPLE: BREAK-EVEN POINT

Facts:

Monthly overhead (G & A) is $50,000 ($600,000 annually)

Average gross profit 10 percent

Solution

Break-even volume is $6,000,000 ($600,000 annual G & A divided by 10 percent gross profit)

When considering alternative profit models, try using the same concept as the break-even calculations, but allow for profit in one of these two ways:

1. Planning profit as a dollar amount: For example, let's say the target is a bottom line profit of $200,000. Add that amount to annual overhead before dividing by gross profit percent to get the target annual revenue. The formula is annual overhead in $ + target profit in $) / gross profit %. See the calculation in the featured example:

EXAMPLE:

PLANNING PROFIT AS A DOLLAR AMOUNT OF REVENUE

Facts:

Monthly overhead (G & A) is $25,000 ($300,000 annually).

Budgeted profit = $200,000, 30 percent gross profit

Solution:

Targeted annual revenue is $1,666,667 ($300,000 annual G & A + $200,000 budgeted profit, divided by 30 percent gross profit)

---

2. Planning profit as a percentage of revenue: For example, let's say the target is a net income of 5 percent (before taxes). Subtract that planned net income percentage from the gross profit percentage before dividing it into the annual overhead. The formula is (annual overhead in $) / (gross profit % – target income %). See the calculation in the featured example:

EXAMPLE:

PLANNING PROFIT AS A PERCENT OF REVENUE

Facts:
Monthly overhead (G & A) is $25,000 ($300,000 annually).
Budgeted profit = 5 percent, 30 percent gross profit

Solution:
Targeted annual revenue is $1,200,000 ($300,000 annual G & A + $200,000, divided by 30 percent gross profit minus budgeted profit of 5 percent).

13.13   RSCORE

The RScore measurement is the result of extensive research by Dr. Tom Schleifer. It is a single formula that determines whether a construction company's financial condition is improving or deteriorating, and to what degree.

Unlike any single ratio discussed here, this formula consists of a combination of interrelated financial and organizational performance indicators. These factors are:

1. Available profit—the portion of profits that can be taken out of the company, or used for growth, without materially affecting the company's financial foundation

2. Real earning power—the product of available profit and the financial turnover of the company's total assets

3. Debt structure—Discussed as leverage in an earlier section, it is measured by the debt to worth ratio

Simply stated, this indicator measures the ability of a company to produce its work at a profit while effectively utilizing its financial resources (essentially keeping debt low in relation to equity). The RScore—and, of course, the company's actual risk—increases when the company's ability to earn a profit is diminished, or when their financial structure becomes more highly leveraged. If both diminished profit earning ability and a highly leveraged financial structure occur at the same time, the risks are multiplied.

The following is the RScore formula:

$$R = [1 - NP/S - (10 \times (TL/(TL + TA)) \times (NP/S))] \times (S/TA) \times (TL/E)$$

Variables:
NP = net profit
S = sales (revenue)

TL = total liabilities
TA = total assets
  E = equity

When the RScore is rising, the company's risk is increasing. Conversely, a falling RScore is good. The changes in this indicator are more significant than the actual value. A significant increase in RScore definitely sends up a red flag well in advance of the company's being in real trouble.

Smaller contractors and those with outside financial strength, like an owner or related company with substantial resources, are able to tolerate higher RScores than larger companies or those companies without access to outside resources. They can tolerate these risk levels since the company's available profit can be supplemented with outside funds.

Certain management actions, particularly increasing discretionary overhead items like owner salaries to reduce taxable income, have a strong negative impact on RScore as well as other indicators.

An easy to use RScore calculator can be accessed free at http://rscore.harddollar.com.

13.14 CHANGE PERCENTAGES

A series of useful gauges can be found in the calculation of change percentages for income statement elements. For example, if revenue increased from $5,000,000 to $6,000,000 in one year, the change was 20 percent ($1,000,000 additional revenue divided by $5,000,000 base year revenue). Consistently high revenue growth percentages usually lead to deterioration in the balance sheet (the company's financial health), so this indicator is useful in setting expectations and identifying or confirming causes.

It is also helpful to calculate change percentages for other income statement elements like gross profit, G & A, and net income, and compare those to the amount of change in revenue to help identify issues. For example, noting that revenue increased 12 percent while G & A increased 20 percent will help spot an issue that is harder to see by just looking at pure numbers in the income statement.

13.15 SUMMARY

It should be obvious that developing skills at financial analysis is demanding and important. Thanks to spreadsheets, the actual calculations cause little difficulty; however, the key to success is the art of interpreting those indicators. Like most art, analysis requires persistence, patience, and practice (especially to become a master). Talent, a natural ability to interpret financial information, is nice, but anyone can do quite nicely with practice and hard work.

There are a myriad of possible indicators and interpretations, so it's important to focus on the key issues in a particular company. Ultimately, the trick is to be able to use indicators to spot challenges and dig deeper to separate the indicators to get to the root cause and take appropriate actions.

The steps bear repeating:

1. Determine an appropriate set of indicators for a company that include the four key areas: profitability, liquidity, leverage, and capacity. Include the essential ratios in the following list and consider additional indicators if additional information is required to solve a problem.

   a. Profitability

      i. Essential—gross profit percent, G & A percent, net income percent

      ii. Consider—unallocated indirect cost percent; gross profit percent broken down by completed contracts, contracts in progress, and divisions or work types; return on equity; return on assets

   b. Liquidity

      i. Essential—current and quick ratios, receivables to payables, working capital

      ii. Consider—days of cash, days of receivables

   c. Leverage

      i. Essential—debt to worth

      ii. Consider—debt to revenue, bank debt to revenue

   d. Capacity

      i. Essential—revenue to working capital, annual revenue growth percent

      ii. Consider—revenue to equity, break-even point

2. Have the indicators calculated on a regular basis, at least quarterly (monthly is advisable) and present them in an easy to follow format.

3. Gather any comparable data on indicators available and compare with historical trends in the company and any industry comparable data. Use the standards quoted in the previous sections as key indicators as a starting point. Be sure to watch indicators that the company's creditors have established as covenants.

4. Once there is a report containing the company's trends and relevant industry standards for the key indicators, practice summarizing the information by answering as many of the following questions as possible:

   a. What direction is revenue moving (growing, shrinking or staying level)?

   b. How fast is the company moving in that direction?

   c. How is the company's profitability at the field level changing (up, down, flat)?

d. How is the company's bottom line changing (up, down, flat)?

e. Based on the changes in revenue and profitability, how would the company's health be expected to change (improving, deteriorating, no real change)?

f. How is liquidity changing (improving, deteriorating, no real change)?

g. How does liquidity compare to the standards (above, below, meets them)?

h. How is leverage changing (improving, deteriorating, no real change)?

i. How does leverage compare to the standards (above, below, meets them)?

j. How does volume during the period compare to working capital (too much, too little, about right)?

k. How does current backlog compare to working capital (too much, too little, about right)?

l. Summarize the most important financial conditions in a few phrases.

This may seem challenging, but it is the essence of financial analysis and definitely worth the effort.

CHAPTER REVIEW QUESTIONS

1. Your company has grown this year, and your surety has warned you that you need to improve working capital on your year-end statement in order to continue the growth. What are some productive steps to do that?
   a. Don't buy fixed assets that are not a necessity.
   b. Don't pay debt off too early.
   c. Protect against "junky" assets.
   d. Be realistic with your job profits.
   e. All of the above

2. The definition of working capital is?
   a. Current assets minus current liabilities
   b. Current liabilities minus current assets
   c. Cash plus receivables divided by current liabilities
   d. Assets divided by liabilities

3. What are two common financial capacity measurements?
   a. revenue divided by working capital
   b. revenue divided by assets
   c. revenue divided by total shareholders' or members' equity
   d. revenue divided by cash

4. How is a company's break-even point calculated?
   a. By dividing annual G & A by last year's profit
   b. By dividing projected annual G & A by expected gross profit percentage
   c. By averaging the last two years' profit
   d. None of the above
5. Which of the following is not a common liquidity ratio?
   a. Current ratio
   b. Quick ratio
   c. Receivables to payables ratios
   d. Profit ratio

CRITICAL THINKING AND DISCUSSION QUESTIONS

1. How many years of financial information are needed to do financial analysis and why?
2. Explain the concept of working capital and why it is important.
3. Why is backlog important and how is it calculated?
4. Is it important to target annual volume and why?
5. Explain how you would go about calculating financial capacity and why it is important.

NOTES

1. Preve, L., and V. Sarria-Allende (2010). *Working Capital Management* (pp. 26–41). New York, NY: Oxford University Press USA.

14

PROJECTION AND BUDGETS

A great evolution in the financial awareness of construction company managers occurs when they begin to use financial information to look into the company's future instead of the past.

When this change occurs, the entire financial process in the company improves dramatically as it becomes more important to everyone, not just the accountants, to get it right. Accounting becomes relevant, even interesting, and managers stop looking in the rear-view mirror to see what happened last year or month and begin to concern themselves with what's happening right now and in this coming year (or month).

Up until this point, the skills and processes necessary to make historical financial information timely and relevant have been explored; however, this information can also be used for projecting future financial information. There is reluctance to attempt to create future financial information when most contend they don't even know what's going to happen tomorrow, and, if they did put a projection together, what would they do with it except look at it at the end of next year to see if they were correct. The way to overcome this obstacle is to understand the proper use of future financial information and the process used to create it.

Contractors are usually very talented at projecting financial information. While they may be more comfortable projecting financial information for jobs than putting together a projection for a company, the skills are the same.

14.1 TERMS

"*Projection*" is a financial tool that is used to explore alternative versions of the company's future in numbers. Projections can be built on a profit basis (projecting an income statement—revenue, costs, expenses, and profit), and then this profit projection can be converted to a cash flow projection. Projections are often done under a variety of "what if" scenarios such as best case, worst case, and most likely case. Projections can (and should) be used to test various ideas and plans on paper before actually implementing them.

"*Forecasts*" are the likely versions of projections. The forecast is essentially a prediction and can be used to either predict profit or cash flow.

"*Budgets*" are a refined form of a forecast, used for control purposes by establishing specific financial goals that people will work to achieve.

14.2 THE PROJECTION PROCESS

The projection process typically starts with building profit projections under a variety of conditions—different volume levels, various potential gross profit margins and overhead options—until a likely, realistic, and acceptable scenario is chosen. That likely projection becomes the foundation for next year's forecast. The forecast is then detailed out and presented as a budget that serves as a road map, to which people are expected to conform for achieving the company's profit model in the next year.

The profit budget is then converted to a cash flow budget in order to understand how the intended actions affect the company's cash flow, keeping in mind that it is possible to earn profits, even large profits, and still run out of cash. The cash flow budget is critical for determining the feasibility of future capital expenditures and necessary financing arrangements and is also a final reality check on a plan.[1] As an example consider a projection showing a significant revenue increase for next year, with the increase primarily coming from a new type of work much in demand and very profitable that the company is going into. Even in the first year of this new venture, the expectation is the new type of work will be profitable and any additional incremental overhead to be minimal; therefore, the forecasted profit is excellent. With big profits in the first year and more to come in the future, the new venture seems like a go.

However, when the cash flow projections are made, there are some apparent challenges. It may take a significant investment in equipment to enter this market. Though the overall overhead doesn't rise dramatically for the year, the additions to overhead may all come at the front end of the endeavor. Collections of receivables may be much slower in this type of work, so financing will be required for several months of work before achieving positive cash flow. An inventory of specialized materials that is necessary, or any number of other factors, may stress the company's cash position.

These cash flow challenges can all be overcome with good management decisions, but not if we don't know they're coming. That's why it's critical to look at cash flow projections along with profit projections. Besides understanding what it takes to run a profitable operation, whether it's a new venture or not, cash flow information is critical to implementing a plan well, to making good decisions, and not surprising key creditors like the bank and the bonding company.

> . . . *it's critical to look at cash flow projections along with profit projections.*

Some potentially profitable opportunities may exist in the market because they create cash flow challenges; otherwise, competitors may have already taken advantage of them. With good planning, an apparent cash flow challenge may be used to gain an advantage.

Consider a medium-size company with the opportunity to take on a single project larger than their entire volume the previous year. The project looked incredibly profitable and could lead to additional work with the client and others like them, and management believed they had the manpower and expertise to build the job successfully. However, the time frame to build this project was extremely, perhaps impossibly, tight, and the amount of labor required was far greater than the company's current labor base. The client was notorious for being slow to pay, which was part of the reason why this particular project had such a high potential margin. The risks associated with this example are discussed in more detail in previous chapters and should be reviewed before taking on a similar endeavor.

The company went to their banker and presented the situation, which took a fair amount of discussion, but their banker agreed to temporarily (for the duration of the job) increase their line of credit. The company got the job, and it worked out pretty much as they projected with cash upside down at the low point, but they had adequate credit to get through that challenge, and cash flow turned positive (and then some) a little more quickly than they anticipated. In the end, this job made a huge profit and put this company into a new level in terms of the size and types of work they could pursue.

If the company had passed on this project, it is likely that they'd still be in the same competitive position they were always in. If they took the job without doing projections and just hoped for the best, they'd probably be gone.[2] Thus, if time isn't taken to chart any course, or attempt to predict what is going to happen in the market and within the company, it will play a defensive position and always look backward.

14.3 THE PRE-PROJECTION PROCESS

Before initiating the projection process, a solid understanding of several key conditions is necessary:

- *Financial capacity*—the financial condition (health) of the company, including all key financial analysis areas, as well as the company's status with key creditors
- *Operational capacity*—what the company can do, should do, and can't do in an operational sense, including determining the best type of work, key customers, strength of field staff and estimating, and a sense of the company's capacity for getting and doing work
- *Market conditions*—the availability of good work by type, the strength of the competition, significant new opportunities, and other conditions that help view the market into the future

If a complete and accurate understanding regarding these three conditions is known, the best course for the company could be easily charted. If each of the three factors is variable, the company will take on a lot of unnecessary risk. To increase the company's odds of success and reduce risk, the best strategy is to make the first two conditions known, leaving only the third factor—market conditions—as a significant variable. Methods to make the financial condition of a company known have already been discussed. The third factor, understanding market conditions, is clearly the most challenging and variable of the three conditions and becomes the basis for the "what if" scenarios to build projections.

14.4 KEY OPERATIONAL FACTORS

The analysis of a company's operational capacity can be a part of a strategic planning process, though it doesn't have to be. Terminology in strategic planning varies from company to company, but generally, companies establish their big picture—a mission and a vision—and then establish long-term goals for typically three to five years. The planning process cascades down to the objectives, strategies, tactics, and actions necessary to achieve those goals. Before building a strategic plan, companies often perform an analysis of internal and external factors. This analysis process is sometimes abbreviated as SWOT (strengths, weaknesses, opportunities and threats).[3] As a part of the analysis of internal factors, the company's greatest and unique strength is identified as their "sustainable competitive advantage." This is any advantage that the company possesses that competitors don't have and can't easily duplicate. In attempting to identify a sustainable competitive advantage, beware of general statements such as "we have the best people or subs." Technical factors such as management and production processes, specialized expertise in particular markets or types of work, production capacity, and skills are more likely candidates for achieving a sustainable competitive advantage.

> As a part of the analysis of internal factors, the company's greatest and unique strength is identified as their "sustainable competitive advantage."

A full-scale version of the strategic planning process can be a profoundly beneficial experience for a company, and many companies use it effectively. However, some find it to be too complex and time consuming. Though they often find the planning session itself to be rewarding, companies frequently bog down in the implementation phase of the plan. Many other contractors appear to avoid the planning process entirely due to these perceived challenges.

If the traditional, full-scale strategic planning process doesn't fit the company, there are critical parts of the process that should be undertaken. There are some steps to take to get key benefits of the planning process in a scaled-down format.

With key people in the organization, brainstorm the following concepts:

1. The strengths of the company, including any potential sustainable competitive advantages.
2. The qualities of work (classified by type) that make it good work. Break down the types of work performed by the company into three categories:

a. Best work (A)—work that the company will actively and exclusively pursue

b. Secondary work (B)—work that the company will do if it walks in the door but won't necessarily seek it out

c. Undesirable work (C)—work that the company won't do and will turn it down even if it just walks through the door

3. The qualities of customers that make them good customers. Break the company's customers or groups of customers down into three categories:

a. Target group (A)—customers that the company will actively and exclusively pursue

b. Acceptable customers (B)—customers that the company won't necessarily seek out but will work for them under the right conditions

c. Undesirable customers (C)—customers the company will not work for

Following are some guidelines to build or reject scenarios with some simple concepts:

1. Develop plans and strategies that move toward doing more of type "A" work for type "A" customers, especially using any sustainable competitive advantage.

2. Work on secondary strategies like doing "A" work for "B" customers, or doing "B" work for "A" customers. Performing "B" work for "B" customers is a less than effective strategy.

3. Develop plans and strategies to turn "B" work into "A" work; attempt to change whatever factors are undesirable about it. If the undesirable factors can't be overcome, the "B" work will likely become "C" work.

4. Develop plans and strategies to turn "B" customers into "A" customers; attempt to alter the undesirable factors about them; otherwise, these customers are likely to become C's.

5. Don't do "C" work or work for "C" customers—just stop. The capacity to do good work for good customers will be increased and many problems, frustrations, and low-margin work will disappear.

These concepts are critical in determining an effective profit model for a company and in developing projections.

14.5 THE PROJECTION PROCESS

The objective of the projection process is to play out various market (external) scenarios in projections, using the knowledge about the internal financial and operational capacities as a guide to the best solutions for a company (Figure 14.1). The market scenarios are variations of projected volume and the related gross profit percentage, the first two elements of the profit model.

Figure 14.1   The Projection Process Requires Input from Multiple Areas within the Organization.

Explore three scenarios: optimistic, pessimistic, and realistic. The realistic one is usually selected in the end, but the other two scenarios provide valuable options and insight into a plan. For example, some conservative moves in the pessimistic case can be considered in the realistic case, or at least held in reserve if things go worse than expected. Determining contingent strategies in advance allows for a lot less reactive steps when the chips are down.

The projection process using different scenarios works best if practiced with the help of spreadsheets. Initially, the broad details should be sketched, and then more detail can be added with each analysis. It can be challenging to look at the big picture, when this may be the first time multiple details are considered at once (such as office expenses, bills, etc.). Spend more time considering the big-picture concepts of the company's profit model than thinking about how much office expense or phone bills have been running.

The first step is to prepare a projected income statement, then convert it to a cash flow projection:

1. For revenue, consider recent volume trends, backlog entering the year, likely projects that are on the horizon, seasonality (if volume comes at certain times of the year), and any other relevant information to estimate an annual volume, then spread that amount over each month of the year.

2. Determine both the amount and the breakdown of construction costs by type, labor, materials, etc. The breakdown of costs is needed to convert the profit projection to cash flow, since different costs, especially labor, are paid on different cycles.

3. For G & A, keep it simple for the first round and just come up with a monthly amount. Consider where overhead has been, any known changes, plus any significant changes that will be necessary to achieve the scenario selected in determining revenue and cost (like adding more office space or staff).

4. For other income and expense items, recent averages usually will work, and then consider significant changes based on the scenario, especially if the scenario involves incurring significant debt and a corresponding increase in interest expense.

At this point step back and see if the profit projection makes sense. Specifically, consider the backlog and how it flows into the first few months of the projection. Do the profit percentages and cost breakdowns make sense with what is known about that work?

The near-term periods should be much more predictable than periods farther out, so spend more time brushing in the details on the periods that can be reasonably foreseen. Most construction companies are so volatile that projecting profits for periods like three or five years is often not useful unless they're evaluating a big alternative like a new branch office, or a new large facility, and even then it's usually vague.

People often make their projections too optimistic by just changing key facts in a minor way. For example, the company is averaging $1,000,000 in sales per month in the current year, so they put in $1,100,000 per month for next year because they feel good about next year. Gross profit has been averaging 10 percent, but the work next year should be better, so they use 11 percent; and doing that extra work won't take any extra overhead. As a matter of fact, there is some waste in the G & A, so they project a reduction in overhead from $80,000 per month to $75,000.

> *People often make their projections too optimistic by just changing key facts in a minor way.*

The result is a projected increase of operating income from the $240,000 (2.0 percent of revenue) expected to be achieved in the current year to the $552,000 (4.2 percent of revenue) projected for next year. However, to achieve this, volume must be increased by 10 percent, while increasing field productivity significantly and cutting overhead. Unfortunately, this type of scenario occurs more often on paper than it does in real life. The reality is that increasing volume puts downward pressure on gross profit and upward pressure on overhead. It is possible to defy those trends if a realistic plan is established.

The next step is to convert the projection to a cash flow format. The amount of detail and sophistication required to do this conversion depends on how much variation and detail there is in the profit projection. If the profit projection contains rapid growth or large fluctuations in revenue (for example, if the company is highly seasonal), it will take time to convert revenue to "cash collected" and construction costs to "cash paid out."

The final step is to layer in the balance sheet transactions that affect cash. These transactions fall into three categories:

1. Assets—buying and selling of fixed assets (property and equipment) or other long-term assets
2. Debt—adding new debt and paying off debt
3. Equity—taking out distributions (dividends) or adding new capital (this one is rare)

Include these transactions in the projection in two groups: those that bring in cash (selling assets, adding debt, or capital) and those that use cash (buying assets, paying off debt, equity distributions). If assessing various scenarios, these balance sheet transactions are often very significant and may be the deciding factor.

Table 14.1 is a very simple profit projection. Each of the months contains the same revenue, costs, and G & A. Note that there is one month that contains an expected refund, and that there isn't any detail in the G & A section. This is to first see if the projection works out before brushing in the details in the G & A.

The next worksheet (Table 14.2) is for converting the profit projection into a cash flow format. In the revenue section of the cash flow worksheet, make assumptions regarding collection periods, which cause each month's revenue to be spread into subsequent months. In this abbreviated projection, the amounts collected don't agree with the total revenue for each month, assuming that the retentions aren't collected during this six-month projection period.

In the expense section of the projection are assumptions as to the breakdown of the job costs by type (what percentages of the total cost for each job goes to labor, materials, subs, and other costs), and then how each of these cost types will be paid. Also assuming that labor is paid in the current month, materials are paid in 30 days (the following month), and subs are paid when the company is paid (and retention is held from them). G & A is assumed to be paid equally each month.

Notice that the balance sheet from the month prior to the projection comes into play in the revenue and expense sections, as collections of the prior month's receivables and payments of the prior month's payables and accruals are projected. Below the expense section of the cash flow worksheet are spaces to layer in the balance sheet transactions. The only items that are currently displayed in these sections are the collection of the refund (mentioned earlier), an asset purchase (for cash), and the principal payments on long-term debt (that was already in place before the projection). At the very bottom of the cash flow

Table 14.1 Sample Contractor Income Statement Projection

	JAN	FEB	MAR	APR	MAY	JUN
INCOME STATEMENT PROJECTION						
REVENUE	1,461,688	1,461,688	1,461,688	1,461,688	1,461,688	1,461,688
COST OF CONSTRUCTION	1,322,828	1,322,828	1,322,828	1,322,828	1,322,828	1,322,828
	138,860	138,860	138,860	138,860	138,860	138,860
EXPENSES:						
COMPENSATION						
DEPRECIATION						
ADVERTISING & PROMOTION						
AUTO & TRUCK						
COMPUTER LEASE						
INSURANCE & BONDS						
LICENSES & FEES						
OFFICE EXPENSE						
PROFESSIONAL FEES						
MEALS & ENTERTAINMENT						
RENT						
REPAIRS & MAINTENANCE						
TRAVEL						
TAXES						
UTILITIES						
COMPUTER EXPENSES						
MISCELLANEOUS						
TOTAL G & A EXPENSES	110,751	110,751	110,751	110,751	110,751	110,751
INCOME FROM OPERATIONS						
OTHER INCOME (EXPS):						
MISC INCOME				20,000		
INTEREST EXPENSE						
NET INCOME	28,109	28,109	28,109	48,109	28,109	28,109

Table 14.2 Sample Contractor Cash Flow Projection

	JAN	FEB	MAR	APR	MAY	JUN
CASH FLOW PROJECTION						
REVENUE:						
PRIOR YEAR RECEIVABLES	723,589	241,196	120,598			
JANUARY		877,013	292,338	146,169		
FEBRUARY			877,013	292,338	146,169	120,598
MARCH				877,013	292,338	146,169
APRIL					877,013	292,338
MAY						877,013
JUNE						
TOTAL REV COLLECTION	723,589	1,118,209	1,289,949	1,315,520	1,315,520	1,436,118
EXPENSES:						
MATERIALS & OTHER	198,424	317,479	317,479	317,479	317,479	317,479
LABOR	0	198,424	198,424	198,424	198,424	198,424
SUBS	110,751	484,155	645,540	726,233	726,233	726,233
G & A EXPENSES	452,084	110,751	110,751	110,751	110,751	110,751
PRIOR YEAR A/P	43,939	91,014	45,507	0	0	45,507
PRIOR YEAR ACCRUALS & TAXES			26,893			
TOTAL EXP PAYMENTS	805,198	1,201,823	1,344,594	1,352,887	1,352,887	1,398,394
ADD:						
OTHER COLLECTIONS				20,000		
LONG-TERM DEBT						
DRAWS-CREDIT LINES						
	0	0	0	20,000	0	0
LESS:						
DEBT REPAYMENTS	7,300	7,300	7,300	7,300	7,300	7,300
CREDIT LINE PAYDOWN					150,000	
ASSET PURCHASES						
	7,300	7,300	7,300	7,300	157,300	7,300
NET CASH IN (OUT)	(88,909)	(90,914)	(61,945)	(24,667)	(194,667)	30,424
CUMULATIVE CASH 84,200	(4,709)	(95,623)	(157,568)	(182,235)	(376,902)	(346,478)

worksheet is the net cash effect for each month and the cumulative cash effect. Note the opening cash balance and the cash balance at the end of the month just prior to the projection.

Mechanically, those are the major steps in building a profit projection, then converting it to a cash flow projection. Before leaving this example, a few things about the results of this exercise are discussed.

The company in this exercise projected a profit model with fairly aggressive growth in volume (about 20 percent), slightly improved gross profit, and a modest increase in overhead. Because the growth is aggressive and starts right at the beginning of the year (it isn't "ramped up"), the effect on cash flow is extremely negative. This is a good example of projecting a profit model that works well as far as profit goes but is obviously very stressful when converted to cash flow. This projected course of action is profitable because the strong growth and improved field performance does a good job of bringing in field profits and stretching out the company's G & A. These actions cause cash pressures because they're paying for labor and materials far ahead of collecting the receivables and way ahead of collecting the retentions.

With a projection like this one, remember it's just a first pass. Further detail is needed because, by the third month, the cash deficit exceeds the open balance on the company's line of credit (so they'd be in real trouble).

The next step is to refine the projection. Use backlog to make revenue more realistic. Verify each of the assumptions regarding G & A, cost breakdowns and collection and payment cycles, asset purchases, and so on. The final step involves changing the ideas rather than refining the projection. If the cumulative cash effect is still too negative for the company to withstand after the refinements, it needs to project different actions. Perhaps they can't grow as quickly as they'd like, and definitely can't pay cash for that asset purchase. If there is already this much volume on the books, they may need to work with their bank to get over the cash flow challenge.

14.6 PUTTING THE PROJECTION TO USE

Once settled on the projection that the company is going to use (actually it has become a forecast now), it can be used as a budget by adding sufficient detail in the expense areas. Most accounting systems will report an income statement with comparisons to budgeted amounts. There are concerns regarding the use of budgets. A budget is supposed to be used for control purposes, but be careful that it doesn't become a weapon that people use against each other (or against the company's best interests). It's primarily a guide to verify that the company is on track to accomplish the profit model, and it needs to be a dynamic tool. Another concern is after the budget process is used for a couple years, it's easy for it to get out of shape because some people may propose a safe budget for their area that they know is achievable. Others may simply take last year's numbers and tweak them a bit. Over the course of a few years, undisciplined behavior may result in a budget that no longer resembles the profit model being pursued.

> *A budget is supposed to be used for control purposes, but be careful that it doesn't become a weapon that people use against each other . . .*

If the company has separate divisions or profit centers, the projections for each of them can be built separately and then assembled. They can usually be input into the accounting system with the divisional breakout so they can be assessed individually.

14.7 SHORT-TERM CASH FLOW PROJECTIONS

Short-term projections (usually three months or less) work differently than the steps described earlier because it's not as necessary to make assumptions about events that haven't yet occurred.[4] The goal is to determine the expected cash effects of events that have already occurred or are well within sight. These short-term projections are normally only used in "cash-stressed" companies, because they take a lot of work (including daily updates). If a short-term cash projection is needed, here are some steps:

1. Set up a spreadsheet for the next three months, broken down by week.
2. Using accounts receivable aging, spread the outstanding balances into the weeks expected to collect them. If expected collections fall later than the periods being worked on, they can be ignored for now and put into the spreadsheet when they're in sight.
3. Using the accounts payable aging, spread the outstanding balances into the weeks in which they are expected to be paid. They are the same thing as collections if they go beyond the time scope.
4. Put in expected payroll by week and payroll-related cash outlays.
5. Using the job schedule, estimate the current month's billings and costs and place them in the weeks expected to collect/pay them. Go into the following months doing the same thing, until the whole period is covered.
6. Put overhead expenses, loan payments, and any other known cash outlays into the weeks in which they are expected to be paid.

Once complete, the next three months are laid out with expected collections and payments. It would be beneficial to keep a record and to maintain a running cash balance, or even bring in a line of credit balance if appropriate.

If the company becomes cash-stressed, a short-term projection will be absolutely critical to survival but don't ignore the long-term view. In a cash flow crunch, if a concerned banker asks for a cash flow projection, provide long-term and short-term formats.

14.8 SUMMARY

To successfully establish a desirable profit model, first go through some pre-projection steps that assess key internal factors (particularly financial and operational capacity), key external factors, and market conditions. Operating in such a high-risk industry requires mitigation of whatever risks are possible in these three areas. The two internal factors, understanding the company's financial and operational capacity, can be easily mapped out using this chapter as a guide. However, market conditions require a more detailed analysis.

After building the projection and checking it against other models, it can be used as a type of road map and guideline for the company. It would be greatly beneficial to keep the projections updated with time and as more accurate information becomes available. Over time, the company will likely become more comfortable with making and using projections and plans for their future success.

CHAPTER REVIEW QUESTIONS

1. What is critical to understand before creating a projection?
   a. Financial capacity
   b. Operational capacity
   c. Market conditions
   d. All of the above

2. The definition of "sustainable competitive advantage" is what?
   a. Any advantage that the company possesses that competitors don't have and can't easily duplicate
   b. Being environmentally conscious
   c. Maintaining lean overhead expenses
   d. The ability to win every job a company pursues

3. Which of the following are steps to creating a useful projection?
   a. Prepare a projected income statement
   b. Convert projected income to a cash flow projection
   c. Layer in the balance sheet transactions that affect cash
   d. All of the above

4. In the projection process what scenarios should be considered?
   a. Optimistic
   b. Pessimistic
   c. Realistic
   d. All of the above

5. How do you establish a desirable profit model?
   a. Assess key internal factors
   b. Assess key external factors
   c. Assess market conditions
   d. All of the above

CRITICAL THINKING AND DISCUSSION QUESTIONS

1. Why should construction organizations go through the effort of doing projections and budgets?
2. Why are key operational factors important to a construction business?
3. How would you convince others to invest the time and effort to make financial projections, instead of just using a wait-and-see approach?
4. Explain the concept of sustainable competitive advantage and why it is important.
5. How are short-term cash flow projections accomplished and what are they used for?

NOTES

1. Park, H. K. (2004). "Cash Flow Forecasting in Construction Project." *KSCE Journal of Civil Engineering* 8(3):265–271.
2. Russell, J. S. (1991). "Contractor Failure: Analysis." *Journal of Performance of Constructed Facilities* 5(3):163–180.
3. Pickton, D. W., and S. Wright (1998). "What's SWOT in Strategic Analysis?" *Strategic Change* 7(2):101–109.
4. Kaka, A. P. (1996). "Towards More Flexible and Accurate Cash Flow Forecasting." *Construction Management and Economics* 14(1):35–44.

15

THE EFFECTIVE
USE OF CREDIT

15.1 INTRODUCTION

Most industries rely on credit. Retailers and manufacturers who make up the lion's share of the companies that are reported about in the news need huge amounts of credit to build facilities, buy inventory and equipment, and so on. They rely on debt, plan on debt, and live with debt. Debt is a big factor in their lives and they accept it.

The fact is contractors need credit (credit isn't necessarily debt) but really don't accept it. In our experience most contractors are uncomfortable with banks in general and don't understand bonding, and they see credit as a necessary evil. Types of asset-based borrowing like mortgages and equipment and vehicle loans are generally more attractive to contractors, and they respect these more than other forms of credit because they get something in return.

Because construction is an inherently risky, highly cyclical industry, being comfortable with significant debt could be a fatal flaw. The process of natural selection may have allowed the debt-averse companies to survive. This chapter discusses the typical contractor's need for credit, how those key credit relationships work, some tips for dealing with key creditors more effectively, and some common questions about credit.

> *Because construction is an inherently risky, highly cyclical industry, being comfortable with significant debt could be a fatal flaw.*

15.2 THE PRIMARY CREDITORS

For most contractors, the two primary creditors are the bank and the bonding company. There are similarities and differences in how each creditor evaluates a construction company and in how they should be approached.

Contractors often don't intend to ever borrow from the bank, but they would like a line of credit while at the same time not wanting the bank to ask too many questions or make them sign too many forms. Often a credit line is set up by a contractor who really does not intend to have a need for it and generally does not want to pay anything for it. Many take a similar approach with a surety.

The biggest obstacle many construction companies face in the credit arena is their attitudes toward the credit grantors, not in the technical analyses or information requests they need to fulfill. It is challenging to get loyalty and respect from key outside credit relationships unless it is given in return.

15.3 BANKING

Often construction companies look to banks to provide them with an open line of credit where they are able to borrow amounts up to a predetermined limit in addition to standard asset loans (Figure 15.1), which are most commonly for vehicles because larger equipment is often financed by manufacturers' credit operations, depository relationships, and money management.

Banks typically prefer approving contractors for asset loans, deposits, and money management services but are not as eager about lines of credit. In some cases they may only be willing to extend a line of credit in order to get the company as a customer for those other services. Contractors cause a lot of money to flow through a bank, which banks like, but unfortunately most of that money doesn't belong to the contractor.

© iStockphoto.com/vladru

Figure 15.1 Banking Relationships Are Critical to Contractor Success.

Most contractors perceive they need a line of credit for two reasons. First, they may hit a few financial bumps that cause a cash flow problem. Second, bonding companies often require a bank line in order to obtain bonding.

Bankers do not look to provide lines of credit to companies to be used as start-up capital, and they do not look to provide financing if a company gets itself in trouble. They are attempting to loan money to proven companies who will do something productive with it, such as buying equipment or getting through a cash crunch when good, collectible receivables are coming in a bit slowly or when positive, sensible, profitable growth has stretched cash flow. In each of these cases, the loans are used to fund something positive: Spend the borrowed money now for productive purposes and collect it plus interest later. Those reasons to borrow make sense to a banker and should be understood by contractors.

When a banker comes across a company that needs to borrow money, they attempt to determine whether the company's situation is positive as described previously or whether the company has developed a problem like they don't have enough work, their work is unprofitable, or their overhead is too high. The success of the loan officer and the bank depends on their ability to determine which group the company is in, will the money be used productively, or is the bank risking loss.

When banks analyze a company's financial condition, they are typically looking for two forms of protection: cash flow and collateral. In order to give the requested line of credit, the loan officer typically needs to demonstrate to the bank's credit committee that a company has the consistent and predictable cash flow necessary to repay the full amount of the line of credit should they happen to draw it all out. It also must be shown that the company has the available collateral to cover the line amount should the cash flow option not work out.

Here are the challenges banks often have in analyzing cash flow:

- Many bankers do not fully understand construction company financial statements, especially work in process calculations, and the causes and consequences of underbillings and overbillings.
- The cyclical and volatile nature of the construction industry makes it difficult to predict consistent cash flows.
- Construction companies frequently provide bankers with late or poor-quality information.

Bankers also consider collateral and often face a dilemma in finding adequate collateral:

- A company may have cash now, but won't have any if they are really in trouble.
- Progress billings will dwindle to nothing, and retentions will be even more difficult to convert to cash if the bank tries to collect them (so receivables are questionable as collateral).

- Miscellaneous assets like prepaids, other receivables, and inventory cannot readily be turned into cash, and underbillings are worth nothing.
- Typically, fixed assets like vehicles and equipment are already secured by the bank or another lender in the form of term loans.

A banker's analysis of the adequacy of a construction company's collateral is uncertain or not solid, and the verdict on the consistency of the cash flow is that it is up and down if they even understand the information they were sent. That said, most of the good contractors have solid banking relationships. The key is that bankers look for contractors who understand and use financial information, are impressed if there are projections, have a sense of where they are going business-wise, and will treat the banker as a valued member of their business team. Bankers obviously also appreciate contractors who have substantial personal assets.

Bankers have an easier time lending if the company always makes money, has solid ratios, provides quick and accurate financial information, and demonstrates an insightful view of business and the bank's role in the contractor's business. The bottom line is that most contractors need a bank to take the time to understand their needs and to become proficient at providing them with the help they need in a professional and timely manner.

> *Bankers have an easier time lending if the company always makes money, has solid ratios, provides quick and accurate financial information, and demonstrates an insightful view of business…*

15.4 BONDING

Bonding companies may seem conservative at times, but it is very uncommon to find a surety underwriter who doesn't truly understand a contractor's financial situation. They know construction, and a contractor must come off positively and professionally to them.

The bonding company's approach is pretty straightforward: They know what it takes to succeed in the construction industry because they have analyzed thousands of statements over many decades, and they want to see a contractor demonstrate those success factors. The conditions that interest the bonding company—financial and operational capacity—have been covered in previous chapters. A company's loss of either one of these capacities will cause the surety to incur a loss.

> *The bonding company's approach is pretty straightforward…*

When a bond is requested for a new project, the surety is concerned that the new project may be that fateful job that the company lacks the ability to complete (perhaps because it is unfamiliar work, is too big, too far away, etc.) and could therefore create a loss. It is also concerned if conditions in the company, unrelated to the new project, have weakened the company (such as the loss of key people, other problem jobs, etc.), exposing the company and the new job with their newly issued bond.

Paradoxically it is the surety's job to be skeptical, and it is their job to support the contractor. Here is the basic approach most sureties use:

1. Develop an understanding of the company's operational strengths and weaknesses and the characteristics of work they perform successfully (type, size, geographic region, etc.). They're much more likely to agree with providing bonds for projects that fit the historical success profile.

2. Develop guidelines regarding the company's financial limits, primarily maximum job size and maximum backlog based on financial capacity.

3. Determine additional risk factors regarding the company's specific situation. For example, does the company have an adequate succession plan, is their information system functioning acceptably, and so on.

4. Based on their perception of the company's risk factors and operational and financial capacities, the surety will accept or reject bond requests.

This is generally how the surety analyzes the financial capacity in more detail. Most sureties focus on working capital, though a few of them use equity as the key factor of analysis. All will assess both of those factors, and they will also use several key ratios including current ratio and debt to worth.

When the surety analyzes a company's working capital, they will disallow (deduct) certain current assets. In most cases, they will disallow all prepaid expenses, 50 percent of inventory, all receivables of over 90 days (not including unbilled retentions) that have not yet been collected, and all unsubstantiated other receivables (particularly shareholder receivables).

Since bonding capacity is essentially the surety's perception of financial and operational capacity, construction companies can improve their bonding capacity in the following ways:

1. Build actual working capital by using the factors previously discussed.

2. Reduce the amounts of any weak current assets; prepaids, inventory, etc. in the balance sheet.

3. Make sure the CPA firm classifies balance sheet items correctly in the annual financial statements.

4. Submit resumes, work histories, etc. that demonstrate that the work you're seeking is well within the company's operational capacity.

5. Demonstrate financial and business awareness by submitting strategic or business plans, a succession plan, projections or budgets, and references to the results of key ratios.

15.5 LEASING

Besides banking and bonding relationships there are credit issues related to leasing:

1. A company can lease virtually anything including things already owned in a sale-leaseback transaction.
2. Like lending, leasing is a form of credit and the pros and cons need to be understood.
3. In an accounting sense, there are two different types of leases—*capital* and *operating*.

The theory behind leasing is that the leasing company will allow a company to obtain the use of an asset over a period of time for which the asset is useful (Figure 15.2). The leasing company can then take back the asset and lease it again, sell it to contractor or someone else, or do whatever else they choose. Leasing enables a company to buy the part of an asset they need and not the whole asset. The company does not need to deal with disposing of the asset or take the financial risk on the *residual value*, the item's value at the end of the lease.

A key factor in deciding whether to lease an asset involves an accounting rule. Some leases are accounted for just like rent, where the monthly lease payment is charged directly to an expense in the income statement and no assets or liabilities are created in the balance sheet. These leases are referred to as

© iStockphoto.com/wesvandinter

Figure 15.2 Lease versus Purchase Decisions Are Complicated And Should Include Input from the Internal Or External Accountant.

operating leases. If the objective in leasing an asset is to protect the balance sheet and working capital, make sure that the lease will be treated as an operating lease by the CPA firm.

> *A key factor in deciding whether to lease an asset involves an accounting rule.*

If a lease is not treated as an operating lease, it is a *capital lease*. In an accounting sense, a capital lease is treated virtually the same as if the company purchased and financed the asset. A fixed asset is added to the balance sheet with a corresponding liability. The monthly payments are charged against the liability and interest expense, like a loan, and the asset is depreciated (technically, amortized—the same thing). In essence this lease acts so much like a loan that it is treated like a loan in the books.

Under existing generally accepted accounting principles (GAAPs), accountants use several tests to determine whether a particular lease will be accounted for as an operating or capital lease. Because these tests require that certain judgments be made, the lease needs to be assessed by the accountants to get an opinion on the accounting treatment. There are also proposals to change these leasing standards in GAAP, so discuss potential leasing issues with an accountant to verify current standards.

It is an obvious capital lease if the title to the asset passes to the company at the end of the lease without paying any residual through a bargain purchase like 10 percent of original cost or $1, which is sometimes seen on office equipment or computer-related items. Another test of a capital lease is whether the company will consume more than 75 percent of the asset's useful life during the lease or whether the total of the payments made effectively pay for virtually all of the asset's original cost. These calculations are the ones that take judgment and are not usually obvious at face value.

Unless a leasing company is specifically structuring a lease to be an operating lease, they very often turn out to be capital leases. A building lease is almost always an operating lease; vehicle and larger equipment leases can be either operating or capital leases, but office equipment and computer hardware and software are almost never operating leases.

Here are some factors to consider when comparing most leases to traditional financing. Not everyone, especially leasing companies, will agree with these. Keep in mind that these factors are generalizations, and there is no standard right or wrong approach. Each situation must be considered on its own merits:

- Cash required at the outset of a lease is often lower than traditional financing; usually first and last month payments down, compared to 10 to 20 percent of the cost of the asset when it is financed.
- It is often easier and quicker to get a lease than a traditional loan, especially for companies that don't have a long or strong credit history.
- The monthly payment for a lease is often lower than the monthly payment using a traditional loan.

- With an operating lease debt can be kept off the balance sheet, preserving working capital and certain ratios, and the accounting is simpler than a loan.
- When the lease period expires just turn it in and get a new one, with new payments of course, which may be good or bad depending on the circumstances and the perspective.
- Financing rates used to calculate lease payments are sometimes higher than the interest rate in a traditional loan.
- It is often more expensive to get out of a lease than a loan. Loans usually have a principal payoff amount (though there is sometimes a prepayment penalty), but a lease is a contract to make a series of payments and there isn't actually a payoff amount.

If the company is very sensitive with regard to working capital or debt to worth and needs to acquire assets, they probably should use operating leases rather than capital leases or loans. With an operating lease, after taking possession of the asset the leasing company is still involved in the use of that asset. A company should be sure they need to and want to use that asset for the entire lease period because they are often difficult and expensive to get out of. The company should also be sure that it can meet any requirements regarding usage or the asset's condition at the end of the lease, or the lease may be more expensive than originally thought.

If the company is not particularly sensitive about working capital, it can consider capital leases or loans when it needs to buy something, or possibly pay cash. Since both of these financing methods have similar accounting treatments, companies usually end up with the asset at the end of the lease. The capital lease needs to be a clear winner on terms or be very convenient to be chosen over the loan because of the lack of flexibility in paying it off if conditions in the company change. There are good uses for leases and loans, so consider the options carefully and get advice from the accountant.

15.6 SUMMARY

Creditors can be split into two groups: those that establish an open credit relationship, like the bonding company and the bank line of credit, and those that provide specific financing (or leasing) for the acquisition of assets. The skills necessary to be successful in building the open credit relationships are much more difficult and comprehensive, and require that the management team understand the banking and bonding processes. Both the bank and the bonding company will tolerate lesser financial and business skills from the management team as long as the credit needs are comparatively small and the financial results are good. If a company executes the business skills well and provides impeccable financial information, both of these creditors will make an extra effort when the company struggles for financial results or needs to expand the relationship. Doing the right thing for creditors is also best for the company.

Here are some typical options for asset-based credit:

1. Pay cash
2. Use traditional financing from a bank, credit company, or the manufacturer's credit organization
3. Lease

The company can decide to pay cash if it has enough working capital and cash after the purchase to do the work they are already doing and, more importantly, the work they *anticipate* doing. When interest rates are low, financing may be a better deal than paying cash, especially if the loans can be paid off without prepayment penalties. If the company's future is uncertain, be cautious and conserve cash.

When a company is choosing between traditional financing and leasing, and the basic terms make either decision feasible (such as the down payment, monthly payment, and loan/lease term), it is important to consider:

1. Whether the company needs to protect the working capital, which favors an operating lease
2. Whether the company wants or needs flexibility in paying off the debt, which favors financing.

CHAPTER REVIEW QUESTIONS

1. Which of the following are not among the challenges that banks often have when analyzing cash flow for a construction company?
   a. Construction companies frequently provide bankers with late or poor quality information.
   b. The cyclical and volatile nature of the construction industry makes it difficult to predict consistent cash flows.
   c. Many bankers do not fully understand construction company financial statements.
   d. Bankers often do not understand implications of the geographic location of the construction company's projects.
   e. None of the above
2. Which of the following are ways that a construction company can improve their bonding capacity?
   a. Reduce the amounts of any weak current assets—prepaids, inventory, etc.—in the balance sheet.
   b. Make sure the CPA firm classifies balance sheet items correctly in the annual financial statements.
   c. Submit resumes, work histories, etc. that demonstrate that the work you're seeking is well within the company's operational capacity.

    d. Demonstrate financial and business awareness by submitting strategic or business plans, a succession plan, projections or budgets, and references to the results of key ratios.

    e. All of the above

3. The construction industry is generally what?
    a. Profitable
    b. High-risk
    c. Cyclical
    d. All of the above

4. An appropriate reason for a construction company to borrow is what?
    a. To cover bonuses
    b. To impress their competitors
    c. To buy necessary equipment
    d. To pay for leasehold improvements

5. Who are generally the primary creditors of a contracting company?
    a. Bank and surety
    b. Bank and subcontractors
    c. Surety and subcontractors
    d. Subcontractors and material suppliers

CRITICAL THINKING AND DISCUSSION QUESTIONS

1. Why do contractors use credit?

2. How do contractors arrange credit, and from whom?

3. Explain what a surety looks for in a construction company and why.

4. Explain the leasing process and why a contractor might use it.

5. What is asset-based credit?

16

MAKING DECISIONS IN VOLATILE CONDITIONS

16.1 THE EFFECTS OF MARKET CYCLES

One of the most elusive management skills in the construction industry is using market cycles and changes in volume as an advantage (Figure 16.1). It can be a lot like learning to juggle while going up and down stairs.

Construction is extremely cyclical; however, the cycles are rarely analyzed to understand their fundamental conditions or symptoms. Another challenge

© iStockphoto.com/Gloszilla

Figure 16.1   Economic Cycles Are Inevitable And Companies Must React to Them.

is that management often cannot react correctly to the cycle the company is faced with because the organization is usually still reacting to the last cycle the company experienced, while simultaneously attempting to anticipate the actions needed to be successful in the next cycle.

For example, if the company has just been through a fast growth stage where good people were hard to come by, management may be hesitant to let go of the productive workers even though the company can't keep them busy. This is especially true if management feels that the normal volume of work is coming back soon.

Similarly, if a company that has been through a long cycle of stagnation or decline and has cut back its overhead severely, when the market rebounds they will be slow to add the overhead back because it was so painful to reduce it. A growth stage often has to last for a considerable length of time before management becomes convinced that growth is going to continue long enough to feel comfortable in building up G & A (general and administrative) again.

> *One of the most elusive management skills in the construction industry is using market cycles and changes in volume as an advantage.*

In a *growth* cycle, which we define as volume growth by 15 percent or more from one period to the next, a construction company will likely go through some or all of the following conditions:

- Higher incidence of "bad" jobs, challenges with customers, and fading margins.
- Cash flow difficulties, including increasing underbillings, problem receivables, and unexpected debt
- Administrative challenges, including slower information, poor quality information, and delayed billings.
- Increasing demands for capital expenditures.
- Human resource challenges, including implementation of problematic compensation systems, high turnover, lack of training, and stress.
- In the early part of a fast growth stage, the net income percentage improves because the company stretched out the overhead over a larger volume of work, but the balance sheet ratios show stress.
- In the latter phases of fast growth, the profitability may fade back to normal (or worse) while the balance sheet ratios may show extreme stress.

In a *decline* stage, where volume falls by 15 percent or more from one period to the next, the company will likely experience some or all of the following conditions:

- In the early phase of a decline, the balance sheet ratios and cash flow improve, but the income statement indicators show stress for one or more of the following reasons:

- When work slows down, the jobs do too; as crews double up or wait for a new job to start, supervision may also double up.
- The estimators feel pressure to *get work*, so they lower margins, use unrealistic productivity figures, and other aggressive assumptions.
- The overhead is too high for the smaller volume, especially when margins shrink.
- Outside creditors react to the income statement stress, get nervous, reduce credit capacity, require more frequent or more demanding reporting, and tighten terms.
- The company struggles with reducing overhead, first with some belt tightening and then with more painful measures.
- Field overhead or indirect costs are too burdensome, and the company struggles to reduce them while still running jobs effectively.

In an extended *stagnation* period, where volume only changes slightly up or down from year to year, the company will likely experience some or all of the following conditions:

1. The balance sheet indicators stay about the same or deteriorate slowly, but the income statement gets stressed over time as field productivity gets lackadaisical and overhead creeps up.
2. There are underutilized resources such as equipment, capital, and people.
3. It becomes hard to justify new equipment or upgrades, salary increases, or improved benefits.
4. Over time, underutilized resources, lack of capital investment, and poorly motivated people result in declining performance in all areas.
5. There is short-term relief from stress in all resources if the stagnation period happens right after a fast growth period.

16.2 G & A STAIR-STEPS

The greatest challenge when a company grows and shrinks in size is managing overhead.

A company's overhead structure doesn't and shouldn't grow or shrink in direct proportion to volume changes. G & A tends to grow and shrink in a stair-step fashion, staying fairly constant throughout a range of revenue. For example, a company needs a certain amount of overhead to productively perform and control a certain amount of work (which varies from organization to organization). However, once a company crosses a certain upper volume threshold that generates too much stress on the organization, the company will need additional overhead. With the additional overhead, the company can now grow to a higher volume level. When the company grows enough to cross another volume threshold, more overhead will be needed and the cycle repeats itself.

This is an example of how this process works when a company crosses a volume threshold that creates enough stress to require more overhead and how doing that affects the bottom line. The company hits $8 million in volume with an overhead of $70,000 per month, gets stressed, and adds additional management, a new computer system, additional accounting staff, office space, and other necessities totaling $20,000 per month. Management is planning to add sufficient revenue to cover the additional expense, but for the moment there is an impact of decreased profit by three percentage points. (The G & A was 10.5 percent of revenue, but company increased G & A to 13.5 percent, so net income decreased by the same amount—3 points).

> *The greatest challenge when a company grows and shrinks in size is managing overhead.*

With the increased overhead structure, the company can effectively perform more work, so management increases volume to $10.3 million and the G & A is back to 10.5 percent of revenue. Then the company continues to increase volume to $12 million with the same overhead, so the G & A is now 9 percent of revenue. This is what is referred to as stretching out overhead. If the company maintains stable gross profits as they grow, the company's bottom line improves dramatically as it stretches the overhead. Unfortunately, the more the company stretches out the overhead (visualize a taut rubber band), the more likely it is to break, and then the company experiences stress in the form of a variety of negative conditions to manage.

It is common for construction enterprises to grow overhead in stair-steps, adding G & A up to a new step as the company crosses a threshold, then moving across the step to higher volumes before hitting the next step. On the left edge of the step while the company is at the lowest volume in the range, profit is the worst, but stress is lowest at this edge. As the company grows volume, moving toward the right side of the step, profits increase but stress increases too (Figure 16.2). The right edge of the step has the highest profit and the highest stress.

Consider this stair-step concept in conjunction with the conditions of market cycles. If a company is declining in volume and starts at the right side or stressed

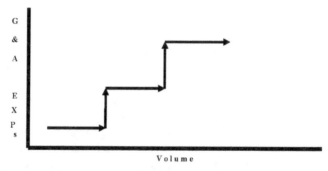

Figure 16.2   As Profits Increase, Stress Also Increases.

side of the step, management should be able to project how far the company can fall in volume to the left side of the step before they have to make a dramatic change in G & A. The more to the left side of the step the company starts, the more quickly management must react and cut overhead. Obviously, management may start belt-tightening as soon as the company moves toward the left side of the step, but companies are not likely to remove whole levels of management, move to a smaller office space, or enact other drastic actions until they absolutely have to.

If a company starts growing rapidly, the same idea holds in reverse: the further the company is to the left side of the step, the longer management has to react as the company moves to the right side. Most companies add incremental overhead enhancements as they move to the right, but few are likely to make substantial additions until they are forced to do so.

The place to stay comfortably is near the center of the step. Unfortunately, that is hard to do. Even if the volume doesn't grow, overhead does, pushing the company back to the left edge of the step. If a company tries to stay close to the right edge of the step, the stress will eventually cause too many breakdowns, and if the company tries to stay at the left side, the high overhead will eat up the profit.

16.3   USING CYCLES POSITIVELY

Using the stair-step concept, it is possible to make positive uses of a market or business cycle in the following ways:

- If the volume is declining, it is a good time to focus efforts on the best customers, the best work by type, and the best people. A decline phase is often a relief to a company if the company cuts out the customers they don't want to work for, the work that doesn't make money, and the people they can do without anyway. Volume declines also can heal up the balance sheet and cash position if the decline phase doesn't last too long.

- Growth is a good stage to stretch out the overhead and to test the limits of effectiveness of the systems and people. The early phases of growth are usually the most profitable of all conditions if the company can manage the cash demands well.

- Stagnation is a good time to sharpen the ax by improving systems and people. It's the best time to train, fix equipment, change systems, and address other project management needs. In a growth phase management is too busy to improve, whereas in a declining market management is afraid to invest money.

All three stages offer value, and if management uses each of them well, the company will benefit greatly.

> *Using the stair-step concept, it is possible to make positive uses of a market or business cycle . . .*

16.4 CONSECUTIVE CYCLES AND FIGHTING TENDENCIES

As mentioned in an earlier comment, management is often unable to react to current market conditions successfully because the company is usually still reacting to the previous market cycle while also looking forward to the next expected cycle. All of these considerations begin to affect the decisions management makes.

For example, if a company has been declining in volume for a while, is still declining, and is expected to continue to decline, the management actions would be clear though challenging. The same clarity would hold true if the company has been growing and expects to continue to grow. Management would be primed and know what to do.

There are two major challenges to overcome:

1. Management gets emotionally stuck in the last phase and does not react quickly enough or strongly enough to the new phase the company is entering.
2. Management can't predict the next phase with any degree of accuracy.

To fight the first tendency, management needs to stay in the present and focus on the management techniques that must be used to take advantage of the different stages. One of the best solutions for dealing with this stuck condition is to have strong outside voices around the company such as an external board of directors or advisors. These should be people management respects, who will be direct with them, and who have enough distance from the company that they can maintain perspective.

To counteract the second tendency, management needs to use the scenario planning discussed earlier to recognize conditions quickly and take appropriate actions. No one can predict the future, but management can envision different *likely* futures, so they are prepared to respond in a timely manner when change occurs.

16.5 SUMMARY

It's challenging enough to get work, do work, and manage a construction company under level, consistent conditions. Unfortunately, reality is harder than that. When market conditions change, managers tend to get stuck in the mode of thinking that was necessary to be successful in the last cycle. Too often, they're complaining about the challenges of the current market and lose the opportunity to use the present conditions to improve the company.

The keys to success are the following:

1. The company knows that it can benefit from all phases of market conditions.
2. Since managers have a tendency to get stuck in the last cycle (everyone does), they must guard against it and use outside advisors who can help them maintain a proper perspective.

3. As conditions change, management needs to change perspective and stay on offense to use the cycle to improve the company.

4. Managers have to be prepared for whichever cycle is coming next by having different scenarios ready.

5. The company has to adapt to the new cycle quickly.

CHAPTER REVIEW QUESTIONS

1. Your company is experiencing a growth cycle. Which of the following is a likely condition that you should watch out for?
   a. Struggles with reducing overhead
   b. Estimators lowering margins, using unrealistic productivity figures, and other aggressive assumptions
   c. Doubling up of crews and/or supervision
   d. Cash flow difficulties, including increasing underbillings, problem receivables, and unexpected debt
   e. Both a and d

2. Your company is in a period of stagnation. Using the stair-step concept, what are ways you can make positive use of this business cycle?
   a. Provide training for employees
   b. Fix equipment
   c. Change and improve systems
   d. Only focus efforts on the best customers, best type of work, or the best people
   e. All except d
   f. All of the above

3. What challenges must be overcome when changing market conditions are encountered?
   a. Management can't predict the next phase with any degree of accuracy.
   b. Management may listen to advice from outside voices such as an external board of directors.
   c. Management gets emotionally stuck in the last phase and does not react quickly enough or strongly enough to the new phase the company is entering.
   d. Both a and c
   e. All of the above

4. G & A tends to grow and shrink in a what?
   a. Straight line
   b. Slight up or down curve
   c. Stair-step fashion
   d. A random pattern depending on the market

5. During declining cycles contractors will do which of the following?
   a. Always lose money
   b. Always make money

    c. Use the cycles to their advantage

    d. None of the above

CRITICAL THINKING AND DISCUSSION QUESTIONS

1. Explain the impact of market cycles.
2. How would you prepare for and address declining market cycles?
3. What occurs in a stagnation cycle and how should a contractor react?
4. What is the stair step concept of G & A?
5. In structuring a construction organization, how would you prepare for various cycles and manage through them?

17

SUCCESS FACTORS FOR A CHANGING INDUSTRY

The next two chapters consider success factors for contractors in the changing dynamics of the construction industry, specifically around how clients and buyers of construction services are seeking to minimize their risks and costs through different procurement methodologies, contractual structures, and operational requirements. For example, though the majority of construction dollars are still awarded to the lowest cost option,[1] there has been consistent growth and an abiding presence of alternative delivery methods[2] within the industry, especially for general contractors. (Most subcontracting trades are still awarded based upon price, regardless of the contract structure of the owner-general contractor agreement.) These delivery systems commonly include design-build (DB), construction manager at risk (CMAR or CM@R), job order contracting (JOC), integrated project delivery (IPD), and several other arrangements or hybrids of the common methods. Of course, other dynamics beyond delivery method are in play as the construction industry and its clients evolve their expectations of the vendor community. With knowledge of how clients are seeking to minimize their risks and costs, a contractor can more effectively compete for contracts and more efficiently work with their clients once the contract has been won. This results in higher margins while still being price competitive for the work, since working with a better client enables a contractor to be more efficient and perform the work with less interruption from the client's efforts and personnel. In some cases, it can also increase the contractor's competitiveness for future work. However, before moving into success factors for these changing dynamics, a perspective of how the industry functions should be explored.

17.1 WHAT A CLIENT WANTS

Most clients who buy construction services must procure and deliver design and construction services within regionally established legal parameters. As such, the specifics of how and in what way a client specifically expresses their desired design and construction outcome will have various nuances depending upon regulatory requirements. But from a general perspective, all clients are attempting to elicit value from project proposers (or bidders) regardless of regional distinctions.

Value, in its purest sense, is the comparison of one option versus another, considering cost and performance. Clients want the greatest value for their construction dollar and seek to extract this value from the vendor community, first through procurement competition (thus seeking to foster a more aggressive pricing approach); second through contract documents, requirements, specifications, standards, etc.; and third through client management and inspection of the contractor's efforts during the actual construction process. However, the contract is signed before any actual work is done and any project specific, quantifiable performance is available. (Performance is a future condition that is promised as part of the contractor's proposal or bid. In essence a bid or proposal is a statement made by the contractors that for a certain amount of money they promise to do all that they have been asked to do.) Since most clients of construction have had one or multiple bad experiences working with contractors, whether due to cost

> *Value, in its purest sense, is the comparison of one option versus another, considering cost and performance.*

overruns, delays, quality issues, or just an overall unpleasant engagement, they have learned that simply writing performance requirements into a contract does not make it actually true.

At first, this may be a difficult concept to understand, since society largely assumes that contracts provide protection, which they do, but typically only in the minimization of loss. For example, in a marriage a prenuptial agreement only protects a person after the marriage has failed, but the contract does not prevent the marriage from failing or from even being a bad marriage. Within construction it is much the same. For example, liquidated damages only apply once the project is already late. A performance bond is only effective after the poor performance has already been experienced. If it were possible to contractually guarantee performance, clients would never experience poor performance. It is rare to find a contract that says it is okay to finish late or over budget, yet projects regularly do.[3] Furthermore, the penalties that are available for response to low performance are retroactively effective (the penalties can only be activated once the poor performance has already happened), so the use of penalty methods is in and of itself a challenging experience for most clients. As a result, since many owners have realized that performance cannot be regularly predicted and in actuality cannot be contractually guaranteed or mandated, they then further rely upon the quality of the construction documents and their own internal managerial expertise to help minimize the risk of low performance and to achieve value from their construction dollar.

An increased reliance upon client management and the quality of the contract documents has unfortunately been hazardous, if recent historical trends are to be considered. Clients report that the quality of contract documents has regularly been decreasing, along with the level of accountability of the design, architecture, and engineering industry.[4] Moreover, in a competitive bid situation, it may actually be easier for a contractor to achieve a profit on a project if the contract documents (drawings and specifications) are of a lower quality rather than if they are of a higher quality. If the contract documents are good, the pricing will be tighter and there will be less opportunity for change orders (and maybe some extra margin) after the contract has been signed. Clients are aware of this and so much of their effort to increase their value has centered around the various parts of the contract and its delivery methods.

17.2   A CLIENT PERSPECTIVE OF THE CONTRACT

To maintain competitiveness and the ability to fairly evaluate proposals or bids, clients must treat the contract requirements, standards, and specifications as minimums for the solicited level of performance. This is a matter of legal convenience and courtesy to the competitors so they may be eagerly engaged to offer realistic and highly competitive proposals that will be fairly and equally considered. This method affords an "apples-to-apples" comparison, where a final selection can be justified if questioned and the value proposition can be quantified at the time of contract signing (in other words, the client's procurement office can show they got a good deal, thereby proving they add value to the organization) (Figure 17.1). If it were not a competition to the minimum, then contractors would necessarily need to submit a proposal of the maximum

© iStockphoto.com/diego_cervo

Figure 17.1   A Procurement Method That Compares "Apples-to-Apples" Makes It Much Easier for the Client to Show When They Got a Good Deal. It Is the Contractors' Responsibility to Stand Out by Differentiating Themselves from Their Competition.

level of performance they could offer, thus resulting in a difficult situation for a client to maintain a defendable and justifiably unbiased evaluation of the submittals (every competitor's proposal would be for a different level of service or performance). It would be akin to comparing apples-to-oranges-to-pears. A selection based upon preference is difficult to defend in case of a protest, since it is difficult to prove value was received as there is no direct comparison available between proposals. It would also require each member of the client evaluation team to have a significant level of expertise in construction and would be frustrating for proposers who are left to guess which flavor of proposal a client evaluation team would like best. With the use of a minimum expectation as the point of competition, the immediate need of a fair proposal process is most easily achieved.

With that said, the client expects that the information, quality, and details stated in the agreement are the lowest level of value or services they will receive from the construction professional they hire. To the client, it truly is the minimum, although it may even be true that clients may expect to see more than the minimum from what they consider to be a high performance contractor.

Conversely, to a contractor proposing or bidding in a competitive environment, the minimum requirement is typically treated as the maximum level of performance that will be realized on the project. Anything above the minimum is not contractually necessary and the cost of achieving such would be directly taken from the potential profit in the job. Also, to do more than the minimum and price this "extra work" into a proposal would actually decrease a contractor's competitiveness for winning the work. A contractor's greatest risk before the contract is signed is winning the job in the first place. Construction is an "all or nothing" industry, where the winner of a contract gets every contract dollar (and all future dollars on the contract via project change orders, for which there is no competition and the price of the changes is negotiated). Second place and lower finishers do not see any of the contract money and have actually spent significant time and resources in an unsuccessful bid attempt. So to minimize the risk of not winning the contract, it is a contractor's best interest in the short term to compete to the lowest interpretation of the contract's minimum requirements and offer the required level of performance at the lowest prices possible, thus increasing the probability of getting the work.

The divergent perspectives of the client and contractor's expectation of performance as derived from the minimum requirement is a consistent source of consternation and project management difficulty on many projects. The lack of understanding between the client and contractor project teams around how the other views the contractual obligations can cause clients to feel they need to rely even further on their own ability to manage the contractor's work and attempt to extract value by high levels of oversight and involvement on the project. This

This frustration of divergent perspectives exists beyond the project level and pervades both client and contractor organizations.

frustration of divergent perspectives exists beyond the project level and pervades both client and contractor organizations.

17.3 HOW CONTRACTS ARE AWARDED

In general, when clients of reasonable size have a construction need, most of them turn to their procurement or contracting departments to prepare and manage the procurement process for the selection and contracting of the construction professional. A procurement officer is assigned to work with the client user group and project management team (sometimes known as a capital project or facilities management department) to prepare the solicitation, manage its release, collect the subsequent proposals, and enter into an agreement with the successful proposer. From the procurement officer's perspective, their "project" effectively ends once the contract is signed and the ownership of managing the work transfers over to the project manager assigned by the capital projects department.

It is also common to see the contract signing process inefficiently transacted, since it is rare to find an individual who enjoys the tedious and legally encumbered process of contract negotiations. For this reason, a "let's just get through this" attitude can be present. Also, procurement focuses on solicitation and selection; final contracting may or may not fall under the procurement officer's responsibility. Furthermore, the primary drivers of a capital project are typically the client's end users and capital projects team, both of whom desire to get started with the actual design and construction of the final facility. To them, the upfront procurement and legal process can be seen as bureaucratic obstacles that must be navigated through before the "real" work can begin. Indeed, it is not unusual with capital projects to find that the clients' project managers (PMs) feel they know the industry well enough to readily identify the best one or two contractors for a job, thus making the selection process, from their perspective, a potential waste of time. For these and other reasons, a "rush to contract" mentality can pervade, with the primary individuals concerned with contract negotiations seeing their project task as complete once the agreement is signed, and the individuals most interested in the operational objectives of the project seeing the contract as an administrative burden that falls under the consideration of others (Figure 17.2).

The measure of a successful project from a procurement perspective is, as stated earlier, the amount under market (average proposal price) that the project is awarded. It is reasoned that if work is being secured for below average rates, then the procurement department is driving a competitive atmosphere. However, for almost all client organizations, there is no linkage between the evaluation and contract award practices of the procurement department and how the contractor actually performs. It is not atypical for a procurement officer to receive kudos for securing a low contract price, and then for the same project the client's project management group is criticized for not being able to bring it

Figure 17.2 A "Rush to Contract" Mentality Is Often Seen during the Procurement Stage, Which Can Have a Negative Impact on the Construction Phase.

in on time or within budget (or both) while working with the lower performing, "lowest price" contractor that procurement hired at such a great price. The expectations and goals of internal client groups are not typically aligned from a performance measurement standpoint. This misalignment also exists within contractor organizations.

17.4 HOW CONTRACTS ARE WON

In most construction companies, there is an organizational divide between the individuals that get the work and the individuals that do the work. It is often thought that winning work in the construction industry is extraordinarily difficult when compared to other industries. This is not necessarily true. Winning work in construction can in actuality be quite simple; one needs only to bid very low, and the work will quickly come in. What is extraordinarily difficult is to win work and make a profit on that work. As the initial portions of this book attest, making a consistent profit in the construction industry requires diligence, hard work, and sometimes luck. The business development, sales, or estimating departments in a construction organization are responsible to win and secure work for the operations group to execute.

Similar to the client organization, the immediate motivations and performance structure of the individuals tasked to win work are built around contract dollars awarded. The responsibility of these individuals is to make the promises to the client on behalf of the construction firm. These promises include price (or how cheap their firm can do the work compared to other firms), how good the people are that will be working on the project, and how much knowledge the firm has about how to do the specific type of work in question. The individuals that propose and make the promises on behalf of the construction firm are not typically the individuals that are tasked with fulfilling the promises made. Estimators, vice presidents (VPs), and business development professionals make the promises and then work to minimize the specificity of the promises into the contract (while the owner is seeking to maximize the specificity and rigidity of the promises within the contract). With this said, for the estimators, VPs, and sales staff, their "project" ends once the contract is signed and they are off to win the next contract. The responsibility of executing the expectations made and agreed to between the client's procurement team and the construction firm's sales team is left to the project managers (for the client and contractor), superintendents, foremen, and so on.

A frequent lament of project managers is that the greatest risk that they face is how to actually do all that their estimators (or VPs, sales, etc.) promised that they could do. This is an inefficient risk perspective, yet it is one that pervades the industry.

> *A frequent lament of project managers is that the greatest risk that they face is how to actually do all that their estimators (or VPs, sales, etc.) promised...*

17.5 RELATIONSHIPS AND CONTRACT DIVERGENCE

As the project is moved from contract award to actual execution and delivery, the key project team members (client and contractor project teams) move to the forefront. These key people may or may not have been involved in the development of the request for proposal (RFP), proposal response, and contract negotiations. Often the first time a construction project manager and his or her team look at a project is after the award has been made and the contract has been signed. The contractor mantra of "first, worry about getting work, and second, worry about how to actually do the work" rings true for many project managers and superintendents.

As the two project managers (the client's PM and contractor's PM) meet and begin to work together, there can be an unstated objective of "how are we going to actually do this?" Indeed, the beginning of a project is often a jovial and happy affair. It is normal to be in a project kickoff meeting where thoughts and phrases such as "we have a great team here," "I'm looking forward to working with you," "I can't wait to get started," and "this is going to be a great project" fill the air. The client's PM is seeking to ensure that the contractor delivers, and the contractor's PM is seeking to make a profit while keeping the customer satisfied.

A normal behavior pattern is for the two project managers (or key critical team members) to seek to form a relationship, because at a subconscious or perhaps even conscious level they realize that they may need that relationship in order to solve potential future challenges and problems they will encounter on the project. As the contractor's PM has a greater level of personal risk on the project and is the reactive and subordinate player in the relationship, he or she will commonly seek to form the positive relationship through acquiescence and readied compliance to owner requests. In other words, at the beginning of a project the contractor will say "yes" to too much of what the client asks for, because they view the need to build their relationship with the owner's PM as of paramount importance.

Once the relationship is established, the critical team members begin working together to execute the contract expectation, which they legally assume to be understood and interpreted the same by all participants. As the two parties work together, the mentality of many PMs is one of task orientation, a "get things done" approach where a good working relationship facilitates quick decision making and assures continual progress to the accomplishment of a goal. Unfortunately, this approach can at times cause all parties to deviate from the contractual expectations. If a significant disagreement around the understanding or interpretation of the contract expectation does arise, the only recourse is to refer to the agreement from which both parties have potentially deviated (Figure 17.3).

© iStockphoto.com/SharpPhotoPro

Figure 17.3 The Strong Tendency to Rely on Relationships Can Backfire When Contract Terms Are Not Clearly Defined before Beginning the Project.

A contract, once signed, is typically not significantly examined again unless there are problems on the project. These instances are then complicated by that fact that the contract was developed by individuals who did not typically plan or discuss operational details during its negotiation and who are not the ones involved in the day-to-day execution of the work.

17.6 THE CLIENT'S VEXING PROBLEM

In considering a portfolio of projects across years of experience, a core problem that stimulates client tinkering of the construction process is the vexing reality of inconsistency in performance from one project to the next. The lack of consistent performance, despite the use of the same contract methods, documents, terms, approaches, and even people is frustrating to clients of all industry sectors.

In the design-bid-build, low bid form of contracting, which is the most common and oldest method of project delivery, some projects go well, others go average, and some are sources of pain, loss, and agony. All of this is in spite of the construction client's use of the same procurement approach, contract structure, contract documents, and all other regulatory and organizational considerations. In other words, sometimes the low bidder is a good contractor and sometimes the lower bidder is a less qualified contractor. There is no control and little predictability around this from an owner's perspective.

For the past several decades, buyers of construction have been seeking to control more of their projects' destinies by adjusting how they compete, award, and deliver work in hopes of finding the best combination and timing of arranging, contracting, and integrating the key industry participants (owner, designers, contractors, etc.). Simply put, buyers of construction know construction services are not a commodity, and that just looking at price to decide who and how to work with the construction professional is potentially not the best approach. The term "alternative delivery method" refers to the process of adjusting how construction projects are competed and delivered.

17.7 GOALS OF ALTERNATIVE DELIVERY METHODS

Perhaps the oldest of the alternatives to low bid is the delivery method of design-build. In the scenario of a challenging design-bid-build arrangement, the owner is stuck between the architect design team and the general contractor. The distinction between what the drawings infer should be included in the cost of the bid and what errors and omissions should result in change orders is a common point of consternation. The owner can see that the design professional's product may not be to the level that was expected, while also having no effective way to prove that the contractor's cost submittals for change order requests are reasonable. The mediating and lose-lose possibility between the design team and construction team is an uncomfortable and unnecessary position for the client. The contractor is the newest comer to the party, as the design team and client team have been working together for quite some time (years is not uncommon) before the contractor is even considered. Being the third and final member of the owner-designer-contractor team and integrating into the established owner-designer relationship can be difficult. For all of these reasons, and many others, owners developed the concept of design-build delivery.

Design-build delivery seeks to minimize the client's risk around having an inefficient arrangement between the designer and contractor, and looks to increase schedule efficiency by allowing fast tracking of the construction schedule (where construction activities, long lead item purchasing, etc. can commence despite the design not being complete). This is juxtaposed with design-bid-build, where 100 percent complete design is needed before going out to bid.

The perceived inefficiencies of a designer and contractor relationship in design-bid-build are intended to be eliminated by requiring the design and construction professions to be contractually or organizationally obligated to each other before competing on the design-build solicitation. In other words, owners thought that if a designer and contractor can't get along in besign-bid-build, then we will force them to get "married" before we let them compete and execute our work.

One major unanticipated problem this created for clients was now they could no longer accomplish with proof their main objective of eliciting value from project proposers. This is true in that the full contract amount is being signed for (albeit under a variety of different available modes) even before any significant design has been accomplished. Before, in design-bid-build, the smaller design costs were expended with a measurable deliverable before the vastly larger construction contract costs were aggressively competed, thus staggering the value proposition and allowing a high level of competiveness on the largest portion of the total project cost. Under design-build there is not much true price competition as most proposers seek to use the entire available budget, instead of delivering a predetermined scope, priced as competitively as possible.

The lack of price competition, linked with fuzzy performance metrics and questions of whether design-build projects truly and consistently performed better than design-bid-build projects, led to the creation of the construction manager at risk (CM@R or CMAR or "pure CM") delivery method. Though there are different ideas of where CMAR originated, it is likely that it was from a hyper-low-bid model where owners sought to eliminate a general contractor's overhead markup on subcontractor work by contracting directly with the subcontractors. Under this model, the owner would first hire a design professional to develop 100 percent complete construction documents. The owner would also hire a construction management consultant (CM) to assist with scope pricing of the design, constructability, and scheduling, and to take the design and break it up into bid packages that would elicit the best pricing from the industry. The owner would then let multiple bid packages, one or more for each major trade. This is in lieu of the traditional model of one bid package being let with all trades included (where only general contractors would compete on bids directly to the client).

The CM would then identify the best price submitted for each bid package option and arrange the contracts of multiple prime contractors directly to the client organization. Having 15–30 prime contractors signed with the client was not uncommon in the pure CM arrangement. The CMs would then coordinate the efforts of the prime contracts and be paid a consultant fee for their services. Initially the fee was a percentage of the total contract value, but as owners saw the potential conflict of interest between the CM and making bid packages the most cost effective (if the bid packages were inefficient, including items out of the traditional scope of sub areas, etc., then the total project price could increase, resulting in a higher fee for the CM), they moved to a fixed fee agreement. As budget bust still occurred, despite saving all the traditional general contractor makeup on each direct subcontractor dollar, owners sought to make the CM more at risk to bring the project in on budget. This was the advent of the guaranteed maximum price (GMP) contract.

With the greater requirement and risk of ensuring design compliance and cost alignment to the owner's budget, insurance and even bonding would be necessary. A consultant firm is not typically able to cost effectively obtain insurance of the magnitude needed across multiple construction projects of reasonable size, and thus the only potential existing industry player who could facilitate a CM function and meet the financial requirements of a GMP were the general contractors. Pure CM presented an opportunity for another profit stream for general contractors, and as their CMs commonly managed subcontracts in all other traditional design-bid-build work in their portfolio, it was a natural transition for owners to release the many prime contract agreements they were signing under pure CM to be under the general contractor's CM they had hired under a fixed fee, and eventually a GMP with a fixed general contractor fee.

There are many contractual options, fee structures, contingencies, etc. under the GMP, and these will not be explored in this text; nevertheless, the second major alternative delivery method was an evolutionary hybridization of design-bid-build and design-build. The hybrid is one where the construction professional or general contractor can be competed on factors other than price, be brought in to help with cost and construction consideration during the design process, have an open-book contract (where the owner gets to see the subcontractor quotes and ensure the general contractor is selecting the best prices), have fast tracking, and not have many of the potential pitfalls of design-bid-build and design-build.

If clients were satisfied with the value of construction they were receiving in design-bid-build, design-build would not have been developed. Likewise, if these two methods met the expectations of construction buyers, CMAR would not have been developed. And recently, another iteration has been introduced called integrated project delivery (IPD), where even more client involvement, oversight, and integration is required, commonly with shared profit and contracts being involved with a high degree of building information modeling (BIM) and digital representation of the project being executed prior to actual construction.

The client goal, however, remains the same. The construction buyer is still seeking to establish, through contractual terms, a mode of increasing the predictability and consistency of their project outcomes: to have the probability of a "good" job be under relative control and reasonably assured.

The evolution of client sophistication is of critical importance to the current and future success of a contractor. Contractors must be able to document their performance and differentiate their value proposition against their competitors in order to be continually competitive in the ever-changing project delivery modes and contractual schemes. A simple bid proposal to win work will be a diminishing reality, and the ability to prepare and deliver a proposal response to a new type of request for proposals is a skill contractors must develop.

17.8   SUCCESSFUL RFP RESPONSE STRATEGIES

In alternative delivery methods, the ability to differentiate yourself as a construction provider in a written RFP response and in an interview format is a learnable skill that is developed through practice and education. The goal of the RFP is to differentiate the value of your construction company on factors other than just price.

> *. . . the ability to differentiate yourself as a construction provider in a written RFP response and in an interview format is a learnable skill . . .*

Most RFPs request a price or fee proposal (although some U.S. states do not allow price to be considered for public works), various measures of potential future capability of your firm and key team members, and a narrative of how the construction activities will be executed (this may include a schedule or simple bar chart graphic). Estimating construction cost is not an unknown skill for most construction firms and will not be addressed in detail. However, the ability to show and prove the capability to perform at a high level is a skill rarely possessed within the construction industry.

Clients request a variety of items in an RFP that they seek to use in their evaluation to identify the potential future capability of a construction firm on a project in question. Beyond normal financial verification (bonding capacity, insurance, etc.), common items requested include resumes of key team members, past client references that may be contacted, examples of past similar projects, organizational charts of the project team, a narrative statement of qualifications, and a safety record or protocols information, among other things. The qualifications are typically joined with a narrative to the project approach and schedule. It should be remembered that in most solicitations the owner specifies the schedule start and stop dates, and even milestones dates, as part of the submission requirement, so the schedule is often really more about how the activities will be fit in to a predetermined duration that has been decided by the client. Within these criteria, however, it is common for a client evaluation committee member to feel that not much differentiation exists between proposals. Another common issue is that some criteria (like a schedule) where the "best" answer requires a high degree of construction knowledge (which some evaluators will not have), forces the client evaluation committee members to use other de facto criteria (consciously or subconsciously). For example, various studies have shown that other than price, the most common factors evaluators of proposal consider is how the proposal actually looks (formats, colors, layout, organization, nice pictures and graphics, etc.) and if they have had a past experience with the proposing firm before. Humans are naturally comparative, and as such there is no true scale of evaluation but instead a relative evaluation of one proposal against another when determining scoring and preference.

This can be a frustrating reality for a competitive contractor who cannot win work with clients because they have pre-established favorites and the RFP tools they are given to differentiate themselves from their competition are not being used to their full extent.

Performance information, when used properly, provides the clearest differentiation of potential future capability of a contractor. When the performance information is extended beyond the general construction firm and addresses the key project team members, key subcontractors, and other key people, the proposing firm presents a highly differentiated proposition to the client evaluation committee that alleviates their need to find differences.

In other words, successful contractors in the changing construction industry must be able to document their performance and use their performance information to differentiate themselves from their competition (Figure 17.4). Successful contractors must cause evaluation committees to make a decision of who to select for the client through the dominance, clarity, and rigor of their proven past performance.

Moreover, marketing and self-promoting narratives of qualifications and descriptions of project approach must be minimized or eliminated. The clearest method to differentiate an expert contractor from a non-expert or less experienced (or less competent firm) is through the ability to identify, prioritize, and minimize risk. An expert and experienced contractor can talk specifically about risks and solutions to the current project being competed since they have done so many similar projects before. A contractor with less expertise or less experience will not be able to offer specific project risk and solutions and instead can only speak in generalities about solutions or about common risks that are part of every project (such as safety, communication, teamwork, etc.).

To differentiate themselves, contractors must use documented performance information and risk analysis in their RFP responses. These will greatly increase the likelihood of being short-listed to the interview stage. Effective strategies for documenting performance information will be discussed in Chapter 18. The next section of this chapter discusses the development of the written response of an RFP proposal.

17.9 EFFECTIVELY USING RISK ANALYSIS IN A PROPOSAL

Unsubstantiated marketing information decreases the likelihood of a successful evaluation score in the contractor's proposals. Structuring the narrative to include the key technical concerns on the projects within the contractor's

© iStockphoto.com/LuMaxArt

Figure 17.4   Performance Measurements Can Clearly Differentiate You from Other Contractors during the Proposal Process.

solutions is highly effective. Even of greater worth to many clients is to identify the risks that are outside of the contractor's control. These would be the risk issues that could impact the contractor's ability to successfully deliver to the client's expectation, since the contractor does not control them directly. This is a drastic shift from the traditional risk mentality, in which the greatest risk was considered to be the actual accomplishment of meeting the promises made by the estimator and sales staff. In the future, sophisticated clients will require you to think in the client's and in the project's best interest.

A risk description in a narrative response to an RFP should include the description of the risk, an explanation of why it is a risk, and the solution to minimizing the risk. As a contractor, you must remember that the majority of the evaluation committee members may not be highly proficient in construction knowledge. They may have limited experience with building systems, equipment, and perhaps common construction information. An effective proposer explains, in non-technical terms, what a risk is and why it is a risk.

Also, every risk identified should have a solution accompanying it. A risk without a solution is not an effective communication and differentiation strategy. The solution

> *. . . every risk identified should have a solution accompanying it.*

should be non-technical and include performance information, wherever possible, to show the solution is viable. If performance information is not viable for the type of risk in question, then a clear actionable plan of how the risk will be eliminated or mitigated should be given.

Examples of technical and non-technical explanations and solutions of a simple construction risk are featured next.

Consider that a risk or technical concern you have is that a poor roofing system can result in roof leaks, which may inconvenience building occupants, increase complaints, increase maintenance, damage building contents, and be a source of mold issues.

Proposer 1

An inefficient technical tolution to the technical concern:

To minimize this risk, we are proposing a thermally-welded roofing system that has a tensile strength of 2,130 PSI, elongation of 300 percent, tear strength of 312 lbs that has been tested for 10,000, and has a cold brittleness of –30°C.

Proposer 2

A better, non-technical solution to assist the client's evaluators:

To minimize this risk, our proposed roofing system has been installed on over 400 roofs and has had an average roof age of 18 years, in which 99 percent of the roofs don't leak and 100 percent of the end clients are satisfied.

Assuming the two proposers are actually proposing to install the same roofing system, the second example clearly speaks of the concern in simple, non-technical terms and provides performance information around the proposed solution. (Of course, this solution requires the contractor to document the performance of their projects.) To be able to respond in the fashion of Proposer 2 should be a goal of every contractor competing on RFP work under the various alternative delivery methods.

Considering a more specific technical risk, a project existed where the design called for the use of a heating and cooling system for a building based in geothermal technology. Consider the following proposal responses to this risk.

Proposer 1

An ineffective, marketing-based solution:

We will use our twenty years of experience in working with mechanical systems to minimize the risk of the heating and cooling system design.

(continued)

Proposer 2

An effective, plan-based solution to the technical concern:

- We have identified the design of the heating/cooling system as a risk. It has not been used before in this area of the country. We will ensure that the system performance and installation is verified prior to finalization of its purchase.
- We have bid using the best rated mechanical contractor in the area (rated at 9.8 out of 10.0) and the next best available mechanical subcontractor (rated 9.1). (Note that this requires performance measures of subcontractors and will be discussed in Chapter 18.)
- Our mechanical contractor identified modifications to the design to improve output and sustainability of the system that will reduce the cost of the mechanical system by 15 percent. We would be happy to present these details during the interview stage of the proposal process.
- We will ensure that the mechanical system will be provided by one manufacturer, and will be commissioned by the manufacturer, contractor, and general contractor, who will take full responsibility for the commissioning of the system. This cost is included in our proposal.

Another example considers the management of noise during demolition and renovation of a portion of an operational building, so as to minimize its impact in disturbing the occupants.

Proposer 1

A non-solution posing as a solution:

We will work with the user to minimize the impact of noise from demolition.

Proposer 2

A viable solution with performance metrics:

- We have planned to perform demolition during off hours and weekends. This will have a slight impact on our cost (less than 1 percent), but the impact to customer satisfaction justifies this.
- We will also install rubber sheets on the floors to diminish noise and vibrations.
- Both solutions can be performed within your budget.
- Both solutions have been used on multiple previous projects with high levels of customer satisfaction (9.4/10).

When responding, be sure to have a viable solution. A final example is presented next that shows the abilities of an expert versus those of perhaps an average contractor. The risk in this example is around having water at a site to irrigate softball fields that are part of a new city park being built. The grand opening will commence with the mayor welcoming the citizens and the first games of the city softball league being played on the new fields. The site is remote, and no existing utilities are available. Irrigation of the fields to have the grass in place in time for the grand opening is on the critical path. The client and contractor must coordinate with the local water utility company to have water at the site.

Proposer 1

A good, but average plan to minimize the risk, but it does not truly address the real risk:

Coordination with [the water company] is critical. We will coordinate and plan with [the water company] as soon as the award is made to make sure that we get water to the site to irrigate the fields.

Proposer 2

An expert's explanation and solution to the risk that they do not directly control:

- We will coordinate and schedule the water with [the water company]. However, based on past experience there is a high risk they will not meet the schedule (the water company does not meet the agreed and coordinated schedule over 90 percent of the time).
- We will have temporary water lines set up and ready to connect to the nearby fire hydrant to irrigate until [the water company] is ready.
- We will also have water trucks on-site if there are problems with connecting the lines.

In Proposer 2's response, the expertise is in thinking like the client and understanding that simple coordination is not enough based on the past experiences and performance information of working with the water company. Clients in the changing market do not want to hire a reactive, complacent construction company. They want to work with a construction professional that is actively engaged in the project and is thinking in the client's best interests.

A simple response that addresses reality, considers what would be of interest from the client's perspective, and provides clear solutions and performance information (when available) is a highly effective form of proposing. How to develop an effective proposal within your company is presented next.

17.10   HOW TO DEVELOP A WINNING PROPOSAL

This book cannot provide experience or expertise, but it can provide processes and strategies that will help augment your ability as a contractor to differentiate yourself in proposal-driven selection processes and delivery methods. The following process is based on numerous debriefs with winning proposers under alternative delivery methods.

Best Practice Proposal Development:
1. Identify the best available team in your company for the specific type of project being proposed. This could typically include the project manager, superintendent, foreman, estimator, preconstruction manager, key subcontractors, etc., along with the business development team or person responsible to develop the RFP response. Time should be made for these individuals to meet to develop the core of the proposal response (they will not do the actual writing, formatting, etc., but will provide the plan and information the proposal should be written around). Some individuals will need to take a few hours off from the job site to do this.
2. Place the team into a planning room and minimize (or eliminate) their outside distractions. The process may need to be repeated for several days; all key individuals will not be required for all aspects of the planning.
3. Review the RFP and associated documents, and then identify what you know of the client and their specific needs and constraints.
4. Plan out the project in major milestone activities with approximate durations. This can be a simple plan, even drawn on a whiteboard. All considerations must be within the known client requirements, laws, rules, and constraints (such as budget, schedule, etc.).
5. Have the team members think of their past projects that were similar to the proposed one and the parts of those projects where things did not go as well as they had wished. This will help to identify past risks, problems, or challenges they have experienced.
6. Prioritize the agreed risks or concerns.
7. Develop plans to minimize the identified risks, concerns, challenges, technical issues, etc.
8. Within the agreed-to plan or schedule, identify the areas of concern, potential risks that exist, and any other issues that might occur and where they might happen.
9. Within the agreed plan or schedule, identify the areas where information, activity, direction, decisions, etc. will be needed from the client.
10. Next consider what the client has requested, and consider that the client's team may not know what is best for them as they may lack construction or design knowledge (do not assume the client has asked for what is best for them, based upon your past experience of working

with similar projects). Remember, this may be the first time ever, or the first time in a long time, that any member of the client's project team has attempted to deliver a project like the one in question. So as an expert contractor, and in order to more clearly differentiate yourself from your competition, it is requisite of you to be thinking as though you were the client; if this were your money, how would you specify and request the aspects of the project. Identify these considerations and catalogue them.

11. Price out the proposal, giving the client what they have requested. If viable under the procurement rules, prepare pricing for the alternative solutions you develop. Typically ensure that all the costs of managing the risks you have identified are included in your base proposal cost.

12. Allow your proposal team to take this information and prepare the response to the actual proposal, within the requirements stated in the RFP.

This process may seem to take longer than your current proposing process, but it should be more effective. If estimators and business development staff work in a silo separate from the operations and field personnel in proposal development, it reduces the viability and competiveness of a contractor's RFP responses.

17.11 SUCCESSFUL INTERVIEWING STRATEGIES

The interview is typically a critical and highly weighted evaluation criterion within alternative delivery methods, especially in those regions where price is not allowed to be considered. The following best practices are presented after the authors have drawn upon direct experience in thousands of interview processes.

> *The interview is typically a critical and highly weighted evaluation criterion within alternative delivery methods . . .*

Send the right people: Clients are growing acutely aware that most often the people doing the talking in an interview are not the people that will be doing the building. Clients know that there is a reasonable chance that many of the individual contractors sent to an interview they will not see again during the project. The presence of presidents and vice presidents shows the commitment of the company, but they are having decreasing success in differentiating their company from the competition to sophisticated clients. To impress a potential client, send the project manager, superintendent, estimators, and/or other critical project personnel, and let them talk. Of course, they need to know about the project, but sending the actual people that the client will be working with day-in and day-out is an impressive move and provides clear differentiation in the client's eyes (assuming the individuals the contractor sends are indeed experts). The president and vice presidents can still attend, but the key project level personnel should be allowed to do most of the talking (Figure 17.5).

© iStockphoto.com/kupicoo

Figure 17.5 A Successful Interview Should Focus on the Individuals That Will Be Managing the Project rather than Executive-Level Personnel.

What to talk about: First, the interviewees should establish why they are there based on their qualifications. They should let the client know that the reason they were selected by their company for the project in question is because they have the most experience in the company for that type of project or have demonstrated the best ability. They should talk about at least one specific similar project they have worked on before as a team. They should be willing to talk about the good and the bad of the project, how the problems were minimized, and the success that was had. The similar project used should also be part of the performance information that is used for past references as part of the written proposal. The final and most important issue to discuss or present is the project plan (that was developed by your project team as part of the RFP response), with its proposed schedule in simple form (do not show a complex, multi-paged CPM schedule at this point). Identify key risk activities, solutions to the risks, key client activities, and any other pertinent information.

If the client budget is known and is a risk, it should be discussed at this time. If the client schedule is unrealistic and is driving up cost, it should also be discussed at this time. Most client evaluation committees appreciate honesty and straightforwardness. The vast majority of contractors will seek to only say what they think the client wants to hear. As clients and delivery methods evolve, this form of pandering to the client is becoming increasingly ineffective. The value of a project manager and superintendent in an interview is they can speak to their experience and are usually more forthcoming and direct than those of superior rank in the construction organization. Most clients will excuse the perceived lack of "polish" around a good superintendent, because they are seeing the man or woman that they want to work with and are willing to put their faith in for the construction and delivery of their project needs.

The interview team should practice and prepare before the actual interview. For the key people assigned for the first time to this mode of interview or presenting, where real project risk, concerns, solutions, and roles and responsibilities are discussed, it will be a frightening experience. The payoff, if done correctly, is greater competitiveness on projects that have an interview evaluation criterion.

17.12 SUMMARY

Know thy client. The client's efforts will always be centered on value creation, risk mitigation, and cost minimization within the regulatory and political realities of their environment. Client and project expectations of construction contractors are evolving and in the immediate future will require contractors to be proficient in performance justification and documentation. This will include the ability to show past performance, along with the ability to demonstrate expertise within the current project. Expertise is most readily shown at an individual level, using the experience and knowledge of a contractor's key project personnel to address specific project concerns a client should have about the delivery of their project. Success begins with understanding the realities of the environment and then developing a competitive mind-set to win and execute work and understand each client's abilities, needs, constraints, and capacity to effectively facilitate the contractor's plan to optimize the project outcome. The next chapter examines the various uses of performance information and tools that will help construction contractors teach their clients how to be better clients.

CHAPTER REVIEW QUESTIONS

1. Most subcontracting trades are awarded based upon _____, regardless of the contract structure of the owner-general contractor agreement.
   a. Price
   b. Best value
   c. Qualifications
   d. Years of experience
2. Which of the following are frequently encountered challenges within the contract award process?
   a. The client's procurement personnel focus mainly on solicitation and selection.
   b. The client's project management personnel may view the selection process as a bureaucratic obstacle and waste of time.

    c. The legal process of contract negotiations is often viewed as tedious and encumbering.

    d. All of the above

3. Which of the following is not a best practice for developing a winning proposal response in alternative project delivery?

    a. Identify the best available team in your company for the specific project being proposed to develop the project plan

    b. Identify potential project risks in your proposal response and communicate your planned solutions

    c. Place an emphasis on marketing and promotional material to help your company stand out

    d. Identify any considerations that the client may not have asked for that would add additional value

4. Which of the following are key common issues faced by client evaluation committees?

    a. Insufficient differentiation between proposals

    b. Some criteria that may require a high degree of technical construction knowledge

    c. Conscious or subconscious preference for certain proposal aesthetics (such as formats, colors, layout, pictures and graphics)

    d. All of the above

5. To impress a potential client, who should you send to participate in an interview process?

    a. The company president or other senior executives

    b. Key operations personnel (i.e., project manager, site superintendent, estimator, etc.)

    c. Marketing, sales, and business development personnel

    d. General laborers from a similar jobsite

    e. All of the above

CRITICAL THINKING AND DISCUSSION QUESTIONS

1. Describe the pros and cons of the various alternative project delivery methods from the perspective of a proposing contractor.

2. If your company were submitting a proposal to build a new college dormitory, what potential risks would you recommend including in your proposal response?

3. What strategies would you recommend implementing to minimize the negative aspects of the "let's just get through this" attitude that is commonly observed in contract negotiations?

4. Describe how you would go about the process of identifying potential risks and solution strategies for a building renovation project you are proposing.

5. Describe how you would prepare a successful interview strategy with your proposal team.

NOTES

1. American Council of Engineering Companies (ACEC) (2008). *2007–2008 Design & Construction Trends Survey.* Washington, DC.

2. Tulacz, G. J. (2013). "Alternative Delivery Rebounds." *Engineering News Record* 270(16):25–28.

3. Post, N. M. (1998) "Building Teams Get High Marks." *Engineering News Record* 240(19):32–39.

4. FMI / CMAA Fifth Annual Survey of Owners (2004). "Management Consulting— Investment Banking for the Construction Industry." Retrieved September 27, 2006 from http://cmaanet.org/user_images/fmi_owners_survey2004.pdf.

18

PERFORMANCE MEASUREMENT

Simple and effective measurement of performance is rarely seen in the construction industry. For example, most construction buyers do not track contractor or designer performance in terms of project metrics such as dollars, time, change order rates, delay rates, causes of change orders and delays, skills sets, and key individuals. If a client holds an opinion of a contractor's capability to perform, it is typically a general perception based upon either past firsthand interaction with that contractor or what the client has "heard" about a contractor's ability, including word-of-mouth and marketing or branding by the contractor. This opinion of performance is the basis for most client-contractor relationships (outside of a specific project setting). These relationships with a client organization are important to a contractor and their ability to secure work, have a confident revenue stream, and attract new business.

A client-contractor relationship is not sustainable over time and alone is not sufficient to be a reliable business practice. The reason for this is that a construction company does not actually have a "relationship" with any client organization. The actual relationship is between individual people in the construction firm and individual people in the client organization. People are not sustainable. They retire, get new jobs, win the lottery, etc. Many contractors have seemingly gone from having a great relationship with a client to having no relationship with a client. This is not because of a decline in performance, but because the proof of the capability to perform (from the client's perspective) was housed in the minds and opinions of one or a few key people in the client's organization. If these people are no longer part of the client's organization, then the construction contractor has lost evidence of their ability to perform.

> *A client-contractor relationship is not sustainable over time and alone is not sufficient to be a reliable business practice.*

Performance measurement and documentation of proven performance must be housed both qualitatively in the relationships and opinions of individuals and also quantitatively in the use of metrics and performance-based closeout documentation. This chapter explores the uses of performance information by construction contractors.

18.1 WHAT TO MEASURE

The actual metrics that are of significance in the construction industry are simple: money, time, and customer satisfaction.[1] These of course break down into much greater levels of specificity, which will be discussed later in this chapter, but the key to contractor performance metrics is knowing what to measure against. It should be remembered that clients are attempting to elicit value from project proposers, where value, in its purest sense, is the comparison of one option versus another. Thus, for performance information to be of any worth it must be compared against, or to, something else. Clients want the greatest value for their construction dollar and what they set their opinions of value around are the expectations set by the contractors in their proposals (or bids). In simpler terms, performance will be measured against the expectations the contractors set for themselves during the proposing and contracting process.

> ... *performance will be measured against the expectations the contractors set for themselves during the proposing and contracting process.*

18.2 SETTING THE CLIENT'S EXPECTATIONS

The potential disconnect between the contractor's team that gets the work and the team that actually does the work can create risk for a contractor to meet the expectations they have set in the client's mind. Exacerbating this problem is that construction planning, or preplanning, often occurs after the contract has been signed, meaning that the first time the project execution team significantly considers the project and discusses it with the client project team is after the expectations have already been set.

A preplanning technique that has proven to be successful if implemented properly is pre-contract planning and coordination (Figure 18.1). This will require the contractor to suggest to the client's team that before the contract is finalized and signed, the two project teams should be allowed to clarify, detail, and coordinate the construction activities. A centerpiece to this effort must be the identification and minimization planning of project risks, specifically risks that are outside of the control of either or both project teams. By first having the contractor's project team interact effectively with the client's team, expectations can be clarified and, on some occasions, reality can be explained. It is not uncommon for a client's expectation or project requirements to not align with their constraints, specifically the amount of funding available (budget) or time allotted (schedule). Application of preplanning with the client before the contract is finalized is most effective in the aforementioned alternative delivery

© iStockphoto.com/AndrewPopov

Figure 18.1   Pre-Contract Planning Is a Technique to Set Client Expectations And Coordinate Successful Project Delivery.

methods, though it has occasionally been successful in a low-bid environment (but only if the client team is of sufficient capability). Most client project teams will value the opportunity to pre-coordinate with the contractor. Some clients will not see the value in having coordination and planning activities prior to contract signing; if this is the case, a contract should proceed as usual, but the contractor should still strive to incorporate a risk-based approach to the preconstruction activities.

18.3   RISK-BASED PREPLANNING

An expert contractor, able to consistently perform at a high level of competency, will be able to successfully implement risk-based preplanning actions into their projects and negotiations with their participant clients. Best practices around risk-based preplanning (again, ideally completed before the contract is signed) include three major activities: 1) preplanning kickoff; 2) coordination and finalization; and 3) summary review of the agreed plan.[2]

For preplanning kickoff, the contractor should come prepared to present to the client team the specifics of their project plan, ideally the schedule, and the potential risks they see in the project and in meeting the client's expectations. Additionally, if the client is agreeable, the contractor should request the client team to develop a list of concerns or risks they have regarding the project or any questions or concerns they have about the contractor's proposal. These should be gathered prior to the kickoff meeting if possible. Along with the project plan, schedule, and the contractor's and client's risks (presented with proposed initial solutions), the contractor should also prepare and present a draft day-by-day schedule of the preplanning period itself (typically one to three weeks in duration depending on size and complexity of the project) with whom or what the contractor would like to meet, clarify, or coordinate. This is a critical activity as the contractor's plan for the project and its potential risks, to this point in the project timeline, has been created in relative isolation from any client interaction (typically based solely on the information provided in the client's tender documents), and as such, must contain assumptions and uncoordinated, unverified details. As much as possible, the draft day-by-day schedule of the preplanning period should be coordinated before the completion of the kickoff meeting. The takeaways from a properly run preplanning kickoff meeting will be: 1) the client having a high level of clarity around the contractor's approach to the project; 2) an agreed understanding of the pertinent risks to the project, thus setting (or resetting) client expectations for the project and providing clarity around the project constraints such as budget and schedule (and, of course, a contractor's cost should not change from their original proposal); 3) an increase in client comfort with the contractor's ability to deliver value; and 4) the client having a clear path forward on what it needs to accomplish in the forthcoming days to facilitate the contractor's ability to effectively provide the agreed construction services.

> *An expert contractor, able to consistently perform at a high level of competency, will be able to successfully implement risk-based preplanning actions into their projects and negotiations with their participating clients.*

The preplanning efforts themselves should follow the agreed schedule of the preplanning activities. There will be numerous meetings, back and forth discussions, e-mails, phone calls, clarification, etc. Client engineers or technical personnel should be encouraged to participate in addition to designers and other professionals that may be under contract to the client. Coordination and follow-up of who will be doing what and when are critical. Components of the preplanning document must include a detailed schedule that not only outlines construction activities, but includes key client activities (including any key decisions, or when access or information will be available, designer activities, etc.), and any "risky" areas where all parties recognize factors outside of the direct control of the contractor will be at play. (Color coding the schedule has been shown to be an effective communication technique for enhanced understanding of roles and responsibilities.) In addition to the schedule, a risk plan should also be documented to include a prioritized list of the agreed risks, accompanied by risk explanations, occurrence minimization plans, impact minimization plans, and insight into each risk's potential impact to the client expectations. Contractors have used the risk plan to review and include changes that will need to be processed due to design or other errors (if drawings are available).

Some contractors have found that having a separate list of agreed client action items (as well as other parties' action items), detailed with the activity, person responsible for its fulfillment, date due, and contact information of the accountable person, to be a useful preplanning and project management tool. Outside of the contract itself, which includes all the traditional terms and conditions, other information in the preplanning document could include a simplified scope plan, contingency allotment protocols (for both cost and time contingencies), and of course, the agreed to performance measurement and tracking process that will be employed both during the project delivery and at project closeout.

> *Some contractors have found that having a separate list of agreed client action items (as well as other parties' action items) . . . to be a useful preplanning and project management tool.*

It is important to remember that the aforementioned activities are to be conducted by the actual project level personnel (client project manager, construction project manager, superintendent, critical subcontractors, foremen, etc.). The preplanning activities should not be primarily completed or directed by supervisory personnel. Supervisory and executive level personnel should facilitate and enforce the proper completion of the preplanning efforts, providing oversight and review where beneficial.

The final activity of the preplanning should be a summary meeting reviewing the already agreed plan (Figure 18.2). This meeting, unfortunately, is often skipped but a final review to ensure that all key project participants understand the project plan and potential risk items and have clear expectations of each other is an extremely valuable exercise.

An example of effective pre-contract planning was evidenced in a CMAR project for the construction of a $40 million facility. The project site was located in a remote location with uncertain and conflicting municipal oversight as well as unknown water pressure availability. Working in conjunction with the design team (who was hired simultaneously with the CM), the project team identified that the availability of water pressure was a risk and that the available water pressure would be unknown until three or four weeks into the project schedule (excavation was necessary). Depending upon pressure availability, the contractor team (working closely with the design team) identified three potential scenarios as part of the risk expectation: 1) adequate water pressure available, no change in building parameters; 2) inadequate water pressure available, impacting some building parameters, such as quality, and potentially size; and 3) worst-case scenario, no water pressure available, requiring substantial underground work, tanks, etc., which would impact the size and other parameters of the building. A few weeks into the project, the determination identified that it was the worst-case scenario, resulting in a reduction of the project scale. The client's reaction upon being presented this news, while not "happy," was one of acceptance and acknowledgement, along with appreciation of the project team's efforts, and an increase of trust in the project team. If the preplanning activities had not been risk focused, the same risk would have occurred,

© iStockphoto.com/Chagin

Figure 18.2 Once the Plan Is Finalized, a Summary Meeting Ensures That All Key Project Participants Understand the Agreed Plan And Have Clear Expectations.

but it would have been a surprise to the client team, with the possibility of suspicion that the potential concern of water pressure was known (or should have been known) by the construction team and was hidden in order to have an advantage in selection and contract negotiations. Trust would have been diminished.

One of the greatest values of risk-based preplanning is the establishment of expectations that can minimize the element of surprise and the inefficient natural human response. Minimization of the risk itself is preferred, but the secondary effect of minimizing the surprise to the client and the possibility of an emotional response also has tremendous benefit. It is the expectation of the client that performance will always be measured against. With proper preplanning, the ability to perform is increased, in large part to having a more realistic expectation to perform and compare against.

18.4 MEASURING PROJECT PERFORMANCE

At the outset of a typical project, a budget will be known, a duration will be agreed to, and some level of quality, deliverables, or specification will be in place. Measurement of project performance should be built around deviations in cost, time, and client expectations from the agreed plan (made ideally before the contract was signed). Specifically, any time the contractor sees a project risk occurring, or potentially occurring, it should be identified and tracked. The tracking mechanism should include the date the risk was identified, the cause of the risk (owner scope, contractor activity, design issue, etc.), a simple overview of the activities necessary to address the risk and a log of its management, and, of course, the impact (if any) to the cost and critical path of the project. The client's project manager or representative will then rate their satisfaction with how the risk has been minimized. Considering current, new, and potential risks on a weekly basis has proven to be the most effective form of utilization of this form of performance measurement. Typically it will consume 5–15 minutes a week once the construction company has established its risk tracking tools.[3, 4]

> *Measurement of project performance should be built around deviations in cost, time, and client expectation from the agreed plan (made ideally before the contract was signed).*

In addition to tracking risk events with their companion cost and time impacts, the risks themselves should be numbered and correlated back to the project schedule. The project schedule should be tracked against the original plan, where any deviation of the actual schedule or potential future schedule is correlated to the numbered risk(s) that are causing the deviation. This practice should also be employed in tracking the project change orders or modifications. Often a single change order will contain multiple cost items for various agreed to changes. Each of the contractual change orders, in addition to having a cost or time impact, should be correlated to the risk(s) number that tracked the project deviation (from the plan) that resulted in the change. In doing this, an expert contractor will find that the majority of project deviations are caused by decisions made by the client after the contract has been signed. Transparency and sharing of the tracked information can create accountability back into the client organization and has been shown to help clients become better clients to work for and with.

A proper tracking tool, when employed across multiple projects simultaneously and over time, will result in the ability to track the following performance measurements:

- Project performance in terms of percent on budget and percent on time
- Overall customer satisfaction, real time, on any given project and across the construction organization
- Change order rates and totals, with raw and percentage breakouts of the sources of cost changes
- Delay rates and totals, with raw and percentage breakouts of the sources of delay
- Performance information by client, identifying and tracking a client's change order approval rate, difficulty to work with, etc.
- Performance information by project manager with satisfaction, time, and cost performance
- Performance information by superintendent with satisfaction, time, and cost performance

- Performance information by subcontractor, supplier, etc.
- Performance information by designer/architect and specific engineers and firms
- Performance information by project type, size, region, duration, season, complexity, etc.

This information also becomes of extreme worth to proposal development, as a contractor will now have the resources to clearly show the company's performance, along with the performance of the key individuals and team members being proposed. Leveraging a robust performance measurement environment into the proposal process introduced in Chapter 17 will afford a contractor with a competitive advantage in the future as clients become more sophisticated in their selection and contracting processes.

18.5 MEASURING PAST PERFORMANCE

Development of the measurement tools and implementation of the tools will take time and discipline for a contractor to successfully create a consistent performance measurement environment. In the interim, a contractor eager to become more measured and insert this information into project proposals should pursue gathering past performance measurement from past clients.

To be time effective, the information should be qualitatively derived, as any quantitative performance information or project data will most likely be too time consuming and too difficult to obtain effectively. Qualitative performance information should be collected around the contractor's capabilities relating to the management of cost, time, quality, risk, operation within rules and regulation, effectiveness at communication, and overall capability or client satisfaction.[5] These measurements should be taken on a scale of the contractor's choosing, with one-to-five, one-to-seven, and one-to-ten scales being the most prevalent. For example, using a one-to-five scale, a client for a specific past project could rate his or her opinion of the contractor's performance around cost management with a "4." This rating, of course, would be more reflective of the actual performance of the project manager and/or superintendent than that of the construction company itself. As such, it is preferable that for each past performance reference measured, the project information be referenced to not only the firm but also to the key project individuals and key subcontractors, thus allowing a contractor to quickly have a database of past performance for their firm, key team members, and key subcontractors (and suppliers).

18.6 PERFORMANCE-BASED CLIENT RELATIONSHIPS

Maintaining sustainable client relationships entails traditional skills and influence coupled with performance measurement in project documentation and project closeout protocols. Most client organizations will have some form of project closeout (beyond the boxes of project records typically housed in a storage shed or project database or both) that is simple and readily accessible.

The aggregation of the contractor's performance for a client, if positive, can be a highly effective proposal and marketing tool in RFP responses and afford greater trust and flexibility in working with the client organization. Performance-based client relationships are established by enhancing and verifying the interpersonal relationship with the agreed performance documentation. The closeout process, if done properly, is also a positive and relationship-solidifying exercise. It should be formal and scheduled by the contractor. An ad hoc closeout is not as effective in

> *Maintaining sustainable client relationships entails traditional skills and influence coupled with performance measurement in project documentation and project closeout protocols.*

setting the tone and memory that is most beneficial. Some examples of closeout and performance measurement tools are provided in Tables 18.1, 18.2, and 18.3

Table 18.1 is a "top page" or summary of a project closeout; it is from a medium-size contractor and tracks the essential project details. It is built for alternative delivery methods and seeks to capture the essence of the project's story. Table 18.2 provides much more detail around project performance with some key cost and schedule information. Table 18.3 shows some qualitative performance measurement questions. These questions could also be used in the accumulation of past performance measurements. Also, a slight rewording of the questions allows them to be used for individuals and subcontractors (internally or externally), affording the development of a robust performance measurement environment.

18.7 MEASUREMENT AND LEADERSHIP

The use of performance measurements for customer relations, project proposals, and project management requires more than data collection tools, templates, and processes. It requires leadership skills and the discipline to properly use the performance measurements collected. Performance measurements cannot be used to cajole,

> *The availability and transparency of performance measurement across the project team assist in creating the behavioral mindset necessary for sustained risk-based management of a project.*

punish, embarrass, or penalize project level personnel. If the information is used for those purposes, the accuracy, regularity of submission, and usefulness of the information will decrease dramatically. Performance measurement can only be accurately generated from those who deal with a project on a day-to-day basis. For a project supervisor or executive to collect the performance measurement from the project team and then turn around with the same information and communicate to the team that they are ineffective is poor leadership. The team will adapt and no longer provide meaningful information. Performance measurement should only be used to provide positive accountability and transparency, and to assist individuals in the completion of their projects. For example, if a project is behind schedule, proper performance measurement

Table 18.1 Example of a "Top Page" Or Summary of a Project Closeout Report

Project Summary					
Client Name and Project Name					
Delivery Method:		Identify the delivery method here (DBB, DB, CMAR, etc.)			
Owner:		Insert owner org name			
	Contact:	Insert client rep name with contract info			
Constructor		Insert your company name			
	Contact:	Insert your contact name and info			
Designer/Engineer:		Insert firm name			
	Contact:	Insert design rep name with contract info			
Start Date:					
Completion Date:					
Contract Days:					
Contract Amount:					
Final Contract Amount:					
Project Description					
INSERT Short description of the project					
Project Evolution					
(Changes to Project From Inception to Completion)					
Description/Reason		**Cost** Time/Quality/ Money	**Created By** Owner/Contractor/ Engineer	**Owner Satisfaction** 0 (not) - 10 (very)	
Project Team Comments/Quotes					
Owner: Insert any owner comments or info that would be useful					
Contractor: Insert any contractor info or comments that would be useful					
Engineer: Insert any designer/engineer comments that would be useful					

Table 18.2 Example of Detailed Project Performance Measurements

Allocated Funds:	$0	Awarded Duration:	0 days
Proposal Cost:	$0	Original Substantial Completion:	mm/dd/yyyy
Accepted VA Items:	$0	Original Completion Date:	mm/dd/yyyy
Awarded Cost:	$0	Total Project Delays:	0
Total Cost Increases:	$0	Client Delays:	0
Client Increases:	$0	Vendor Delays:	0
Vendor Increases:	$0	Design Delays:	0
Design Increases:	$0	Unforeseen Delays:	0
Unforeseen:	$0		
		Actual Substantial Completion:	mm/dd/yyyy
Final Cost:	$0	Actual Completion Date:	mm/dd/yyyy
Percent Increase:	**0.0%**	Percent Increase:	**0.0%**

will identify the causes, sources of delay, and actions taken thus far to avoid or minimize the impacts. This information may assist in adjusting the current work, provide effective communication and resources, and will certainly be of benefit on future work in terms of planning and personnel assignments.

The availability and transparency of performance measurement across the project team assist in creating the behavioral mind-set necessary for sustained risk-based management of a project. Identification and group recognition of performance slippage facilitates faster response and resolution, both on a micro-scale (project or task level) and a macro-scale (organizational). For example, when confronted with a project problem or concern some project managers will attempt

Table 18.3 Example of Qualitative Measurement Criteria

#	Performance Question	Scale	Rating
1	How well did we manage the project costs?	(1–10)	
2	How well did we maintain the project schedule?	(1–10)	
3	How was our quality of work?	(1–10)	
4	How well did we communicate?	(1–10)	
5	How well did we explain, manage, and document the project risks?	(1–10)	
6	How well did we did we follow the rules, regulations, and requirements?	(1–10)	
7	What is your overall customer satisfaction and comfort level in hiring us again?	(1–10)	

Printed Name (of the client)	Signature (of the client)	Date

to solve the problem before anyone more senior hears of the problem. The problem is hidden by omission of reporting or identification, with the problem not being externally communicated until it is of significant scale to be beyond the project manager's ability to camouflage any longer. A common lament of program managers or supervisory personnel is that by the time they learn of a project problem, it is often a much bigger problem than it needs to be, and if they had been informed earlier, the problem could have been resolved with greater ease. A working performance measurement and reporting process will minimize this; however, to develop a working measurement process requires the organizational leader to develop the institutional trust necessary for the reporting to be beneficially accurate and timely by the project personnel. As project teams see the information helping the organization gain new work, and experience it being used by their supervisors to help the project teams successfully finish their projects, the system will advance and mature quickly. Updates and tweaks will come from the system participants, not just from the administrators. Good leadership creates this environment through steady communication, constant education, consistency, and honesty.

18.8 SUMMARY

Performance measurement can have a direct impact on a construction contractor's bottom line. It helps sustain client relationships, can be used to prove value and differentiate a measured contractor's proposal from a "words only" contractor's proposal, and creates an accountable work environment for internal efforts. As with most of the concepts presented in this book, the ultimate teacher of their implementation will be trial and error, and performance measurement is no different. A contractor's first attempts at becoming more measured, for whatever purpose(s), will most likely not be perfectly correct at the outset. Contractors are encouraged to accept this reality and not try to perfect their measurement tools and processes tools before implementing. Contractors should simply begin measuring, on a small scale, that which is of the greatest value or of the greatest ease. Some measurement is better than no measurement. Developing value-driven performance measurements will be an evolutionary process but one of tremendous worth to the ability of a contractor to be successful in the future of the construction industry.

CHAPTER REVIEW QUESTIONS

1. The actual client-contractor relationship is between which entities?
   a. The construction firm and the client organization as a whole
   b. The construction firm executive and the client organization as a whole
   c. The construction firm as a whole and key project managers in the client organization
   d. Individual people in the construction firm and individual people in the client organization

2. The actual metrics of significance in the construction industry are what?
   a. Cost (budget)
   b. Time (schedule)
   c. Customer satisfaction (quality)
   d. All of the above

3. The purpose of preplanning before the contract is to do what?
   a. Set the client's expectation around the value they will receive
   b. Clarify, detail, and coordinate the construction activities
   c. Identify project risks and develop minimization plans
   d. Extend the duration of contract negotiations
   e. All but d
   f. All of the above

4. Project performance measurements are most effective when they are tracked against what?
   a. Preset performance ranges specified by the client
   b. The original budget, schedule, and quality levels agreed to in the contract
   c. Benchmarks of performance for similar project types and sizes
   d. Benchmarks of historical change order rates in similar geographical regions

5. Recommended leadership practices include the use of performance information to do what?
   a. Identify and penalize low performance
   b. Communicate to project teams that they are being ineffective
   c. Create transparency to assist with the completion of projects
   d. Promote competition between project teams
   e. All of the above

CRITICAL THINKING AND DISCUSSION QUESTIONS

1. If you were a large client organization that purchases lots of construction services, what type of performance information would be most useful for you to know about proposing contractors?

2. Describe some of the underlying conditions that may cause the client's project team to have different expectations of performance than what was stated in the contract.

3. Describe the benefits that can be gained from engaging the client in pre-contract planning activities with an emphasis on risk-based planning.

4. Describe why it is important to have both quantitative and qualitative performance measurements in the construction industry.

5. If you were the leader of a construction firm, how would promote a positive company atmosphere around performance measurement and transparency?

NOTES

1. Egbu, C., D. Kashiwagi, K. Sullivan, and B. Carey (2008). "Identification of the Use and Impact of Performance Information within the Construction Industry." CIB Task Group 61 Summary Report, Amsterdam, Netherlands.

2. Lines, B., K. T. Sullivan, and A. Perrenoud (2013). "Optimizing Cost and Schedule Performance through Best Value Project Delivery: Application within a Design-Build Project." *Journal for the Advancement of Performance Information and Value* 5(1):27–40.

3. Perrenoud, A., and K. T. Sullivan (2013). "Implementing Project Schedule Metrics to Identify the Impact of Delays Correlated with Contractors." *Journal for the Advancement of Performance Information and Value* 5(1):41–49.

4. Sullivan, K., D. Kashiwagi, and N. Chong (2009). "The Influence of an Information Environment on a Construction Organization's Culture: A Case Study." *Advances in Civil Engineering,* Vol. 2009, Article ID 387608, 10 pages.

5. Sullivan, K. and J. Savicky (2010). "Past Performance Information: Analysis of the Optimization of a Performance Evaluation Criteria." *Journal for the Advancement of Performance Information and Value* 2(1):13–22.

APPENDIX

ANSWER KEY FOR CHAPTER REVIEW QUESTIONS

CHAPTER 1

1 (b); 2 (d); 3 (c); 4 (d); 5 (b)

CHAPTER 2

1 (d); 2 (c); 3 (d); 4 (c); 5 (b)

CHAPTER 3

1 (b); 2 (d); 3 (a); 4 (e); 5 (b and d)

CHAPTER 4

1 (a); 2 (d); 3 (b); 4 (e); 5 (c)

CHAPTER 5

1 (d); 2 (d); 3(a); 4 (d); 5 (c)

CHAPTER 6

1 (d); 2 (b); 3 (d); 4 (d); 5 (a)

CHAPTER 7

1 (b); 2 (e); 3 (d); 4 (a and c); 5 (a)

CHAPTER 8

1 (d); 2 (b); 3 (d); 4 (b); 5 (c)

CHAPTER 9

1 (d); 2 (c); 3 (b); 4 (e); 5 (d)

CHAPTER 10

1 (d); 2 (d); 3 (d); 4 (d); 5 (a)

CHAPTER 11

1 (a and c); 2 (d); 3 (b and d); 4 (b); 5 (b)

CHAPTER 12

1 (f); 2 (c); 3 (d); 4 (a and c); 5 (d)

CHAPTER 13

1 (d); 2 (a); 3 (a and c); 4 (b); 5 (d)

CHAPTER 14

1 (c); 2 (a); 3 (d); 4 (d); 5 (d)

CHAPTER 15

1 (d); 2 (e); 3 (d); 4 (c); 5 (a)

CHAPTER 16

1 (d); 2 (e); 3 (d); 4 (c); 5 (c)

CHAPTER 17

1 (a); 2 (d); 3 (c); 4 (d); 5 (b)

CHAPTER 18

1 (d); 2 (d); 3 (e); 4 (b); 5 (c)

INDEX